Studies in Computational Intelligence

Volume 713

Series editor

Janusz Kacprzyk, Polish Academy of Sciences, Warsaw, Poland
e-mail: kacprzyk@ibspan.waw.pl

The series "Studies in Computational Intelligence" (SCI) publishes new developments and advances in the various areas of computational intelligence—quickly and with a high quality. The intent is to cover the theory, applications, and design methods of computational intelligence, as embedded in the fields of engineering, computer science, physics and life sciences, as well as the methodologies behind them. The series contains monographs, lecture notes and edited volumes in computational intelligence spanning the areas of neural networks, connectionist systems, genetic algorithms, evolutionary computation, artificial intelligence, cellular automata, self-organizing systems, soft computing, fuzzy systems, and hybrid intelligent systems. Of particular value to both the contributors and the readership are the short publication timeframe and the worldwide distribution, which enable both wide and rapid dissemination of research output.

More information about this series at http://www.springer.com/series/7092

Brajendra Panda · Sudeep Sharma
Usha Batra
Editors

Innovations in Computational Intelligence

Best Selected Papers of the Third International
Conference on REDSET 2016

 Springer

Editors
Brajendra Panda
Department of Computer Science and
  Computer Engineering
University of Arkansas
Fayetteville, AR
USA

Usha Batra
Department of Computer Science
  Engineering, School of Engineering
GD Goenka University
Gurgaon, Haryana
India

Sudeep Sharma
School of Engineering
GD Goenka University
Gurgaon, Haryana
India

ISSN 1860-949X   ISSN 1860-9503   (electronic)
Studies in Computational Intelligence
ISBN 978-981-13-5166-2   ISBN 978-981-10-4555-4   (eBook)
https://doi.org/10.1007/978-981-10-4555-4

Printed on acid-free paper

This Springer imprint is published by Springer Nature
The registered company is Springer Nature Singapore Pte Ltd.
The registered company address is: 152 Beach Road, #21-01/04 Gateway East, Singapore 189721, Singapore

Preface

Computation leads the path toward a realm of creation and the product that is eventually developed, which is a blend of intelligence, logic, and network. Many decades have been spent in developing the methods and tools of Computational Intelligence such as Artificial Neural Networks, Evolutionary Computation, Fuzzy Logic, Computational Swarm Intelligence, and Artificial Immune Systems. However, their applications are not disseminated as exhaustively as they could be. Computational intelligence can be used to provide solutions to many real-life problems which could be translated into binary languages for computers to process it. These problems spread across different fields such as robotics, bioinformatics, computational biology, including gene expression, cancer classification, protein function prediction, which could be solved using techniques of computational intelligence involving applied mathematics, informatics statistics, computer science, artificial intelligence, cognitive informatics, connectionism, data mining, graphical methods, intelligent agents and intelligent systems, knowledge discovery in data (KDD), machine intelligence, machine learning, natural computing, parallel distributed processing, and pattern recognition. In recent times, it has become more important to study subjects in an interdisciplinary way and provide computationally intelligent solutions to our scientific and engineering ideas. Promotion of this approach is rapidly picking up pace in developed countries where curriculum and research frameworks are designed around this theme and a lot of research has been going on in this field.

Present conference, REDSET (Recent Developments in Science, Engineering and Technology—2016) in association with Springer, is our humble effort in this direction. Promotion of inquiry-based education has been the core competence of the School of Engineering at GD Goenka University since its inception and present conference is yet another step forward. We aim to promote the interdisciplinary nature of scientific enquiry, which leads to engineering efforts to develop scientific and technologically innovative solutions to benefit society as a whole.

We received more than 500 papers from 11 different counties including India, USA., Brunei Darussalam, Bangladesh, Canada, Japan, Nigeria, Gabon, Qatar, United Arab Emirates, and Saudi Arabia. After undergoing a rigorous peer-review process following international standards, the acceptance rate was 35%. Selection was based upon originality of idea, innovation, and relevance to the conference theme and relation to the latest trends in respective domains. The articles include various disciplines of physics, chemistry, mathematics, electronics and electrical engineering, mechanical, civil and environmental engineering, and computer science and engineering. These papers are from scientists, practicing engineers as well as young students from more than 105 universities all over the globe. We hope that the ideas and subsequent discussions presented in the conference will help the global scientific community to drive toward a nationally and globally responsible society. We also hope that young minds will derive inspiration from their elders and contribute toward developing sustainable solutions for the nation and the world as a whole.

This conference proceeding volume contains the contributions presented during the 3rd International Conference on Recent Developments in Science, Engineering and Technology (REDSET-16), which took place at the GD Goenka University campus on 21 and 22 October, 2016. We thank Springer for being our publication partner for this conference. This new book series on "Studies in Computational Intelligence" consists of 21 contributed chapters after rigorous review by the subject matter experts.

Fayetteville, USA   Brajendra Panda
Gurgaon, India   Sudeep Sharma
Gurgaon, India   Usha Batra

Acknowledgements

It is with distinct pleasure that we present to you this book titled Innovations in Computational Intelligence that crowned the REDSET 2016 (Recent Developments in Science, Engineering and Technology) conference, which was held at the GD Goenka University, Sohna, Gurgaon, India, from October 21 to 22, 2016. This book contains some of the best papers presented at the REDSET 2016 and had the clear objective of moving forward Innovations in Computational Intelligence to the next level.

Given that people from every segment of our society, from academia to industry, from government to general public, are inundated with massive amount of information, whether it comes through wired networks or wireless, computers or handheld devices, it is vital that computational methodologies are developed to dissect all that data, filter out noise, and build knowledge that helps in decision making. This aspect of computation has resulted in moving away from the old-fashioned data analysis and stimulated nature-inspired activities including learning, adapting, and evolving computational methodologies. Thus, Computational Intelligence has assumed a major role in Science, Technology, Engineering, and Mathematics (STEM) research and is driving the future of technology. It is our strong belief that this field will continue to mature into an area that is critical to today's high-technology environment and will become a key pillar of our information-dominated society.

At REDSET 2016, we witnessed a highly enthusiastic group of researchers, developers, and practitioners from academia and industry come together to present their research findings on various topics related to Computational Intelligence and its applications. From the glorious success of REDSET, it was evident that this budding conference is evolving into a leading conference in India where eminent researchers from around the world would converge to share their research results with the STEM community. The papers presented at the conference were of high quality with discernable scientific merits and significant research contributions. For this book, we have picked some of best papers that were deemed visionary and cutting edge.

Computational Intelligence is an area that will gain increasing emphasis in the future. The papers presented in this book will provide a treasure trove to academicians, researchers, and information technology practitioners in their quest for scholarship. Thus, we hope the readers will find the book a valuable source of learning and reference for their ongoing and future research and development endeavors.

The glorious success of REDSET 2016 became possible because of the persistent efforts made by program chair, members-program chair, and entire organizing committee. We would like to express sincere gratitude for the constant support made by GD Goenka University Management, hon'ble Vice-Chancellor Prof. (Dr.) Deependra Kumar Jha, Registrar Dr. Nitesh Bansal, and administrator of GD Goenka education city Major Kartikey Sharma. A special thanks to University of Arkansas, for being associated and for participating in REDSET 2016.

We wish to express our sincere appreciation to the members of the technical program committee and the reviewers who helped in the paper selection process. We also thank the authors for their contributions that made us assemble such an exceptional technical set of papers.

Our heartiest and special thanks go to the editorial staff at Springer Ms. Suvira, Ms. Nidhi, and Mr. Praveen Kumar for their outstanding services and support.

Contents

About the Editors

Dr. Brajendra Panda holds a Ph.D. degree in Computer Science from the North Dakota State University (NDSU), Fargo, North Dakota, and is currently a professor at the University of Arkansas in Fayetteville's Department of Computer Science and Computer Engineering (CSCE department). He has been an active member of institutional services at the University of Arkansas, and NDSU. Dr. Panda has served as a member of the program committee as well as general and program co-chair at several international conferences, workshops, and symposiums and has also reviewed papers for journals and international conferences. He holds the Outstanding Researcher Award from the CSCE department, University of Arkansas for the years 2002–2003 and 2008–2009. His research interests include database systems, computer security, computer forensics, and information assurance. He has published a number of influential papers.

Dr. Sudeep Sharma is currently a Professor and Assistant Dean, School of Engineering at GD Goenka University. He holds a doctorate degree in Mechanical Engineering from the Technical University of Clausthal in Germany and M.Tech. in Manufacturing Engineering from the University of Manitoba, Canada. Dr. Sharma has an impressive publication record and extensive experience in the field of engineering and technology as well national and international academic institutes.

Dr. Usha Batra is currently a Professor and Head of Computer Science and Engineering Department at GD Goenka University, India. She holds a Ph.D. in Computer Science Engineering from Banasthali University, Rajasthan, India. Her research interests include software engineering, aspect-oriented programming, enterprise application integration, and health informatics. Prof. Batra has authored many scholarly publications, published in journals and conferences of repute. She is chairing Research and Development Center at GD Goenka University.

Neoteric Iris Acclamation Subtlety

Zafar Sherin and Mehta Deepa

Abstract Strong biometrics is a key element of securing modern-day systems which are characterized specifically by iris. The research study portrayed in this paper deals specifically with the development of a neoteric iris acclamation subtlety. Data capturing marks the beginning process which acquires the biometric sample. This follows with feature extraction which leads to the creation of biometric signature. The developed biometric signature is compared with a particular or several biometric signatures that are being registered in the knowledge database, together designated as biometric templates. They are collected during the enrolment process which corresponds to an identity that is subject verified. When the acquired biometric signature matches with the template, then the identity being claimed is the same identity being stored; otherwise, it belongs to a different identity. The comparisons done between the templates determine the basic distinction betwixt the nodes that are exploited for performing biometric recognition, namely verification and identification. One-to-one match is resulted by the verification process, where the identity of the person is verified by the biometric system. On presenting a new sample to the system, calculation of the difference between the new sample and its corresponding template (which is stored previously in the system) is done and the comparison of the computed difference and predefined threshold takes place. New sample is being accepted if the difference comes out to be smaller; otherwise, rejection of sample occurs.

Keywords Biometric template · Pattern resemblance · Hough transformation

Z. Sherin (✉) · M. Deepa
Department of CSE, SEST, Jamia Hamdard University, Delhi, India
e-mail: Sherin_zafar84@yahoo.com

M. Deepa
e-mail: deepa.mehta12@gmail.com

Z. Sherin · M. Deepa
Department of CSE, FET, MRIU, Haryana, India

© Springer Nature Singapore Pte Ltd. 2018
B. Panda et al. (eds.), *Innovations in Computational Intelligence*, Studies in Computational Intelligence 713, https://doi.org/10.1007/978-981-10-4555-4_1

1

1   Introduction

Figure 1 depicts the processes in biometric system independent of the trait being utilized. Researchers have focused that biometric identification is becoming quite a popular tool and gaining more acceptance in various sectors.

One of the highly accurate and reliable methods to be considered for biometric identification is iris recognition as the iris is considered to be very stable, highly unique and easy to capture when compared with other biometric identifiers. For personal identification, image processing and signal processing, the unique epigenetic patterns of a human iris are employed for extracting information which is encoded to formulate a "biometric template". This biometric template is stored in a database and also utilized for identification purposes.

Figure 2a focuses on view of human eye and its various parts and (b) depicts the headmost prospective of annular component called iris. Iris regulates the extent to which light can insinuate through the pupil, and this function is carried out by the

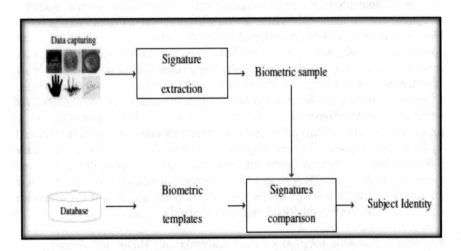

Fig. 1   Processes in biometric system

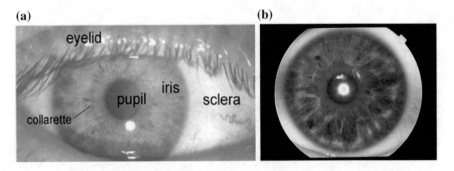

Fig. 2   a Front view of human eye, **b** a view of iris

sphincter and the dilator muscles by modifying pupil size. Iris may have T average diameter of 12 mm, whereas pupil size varies from 10 to 80% of the iris diameter. An iris template can be created from 173 out of relatively 266 peculiar inclinations; hence, iris perception is the most assuring biometric mechanics. The upcoming sections of this paper will describe the various phases of neoteric iris subtlety followed by simulation results, conclusions and future analysis.

2 The Proposed Neoteric Iris Subtlety

The proposed neoteric iris subtlety has been implemented in MATLAB to provide enhanced biometric security solutions. It undergoes the various steps, namely segmentation (iris segmentation/disjuncture), normalization, encoding (template formation or encoding), matching and authentication. The basic operations of the proposed neoteric "crypt-iris-based perception and authentication approach" are specified in Fig. 3 and described below.

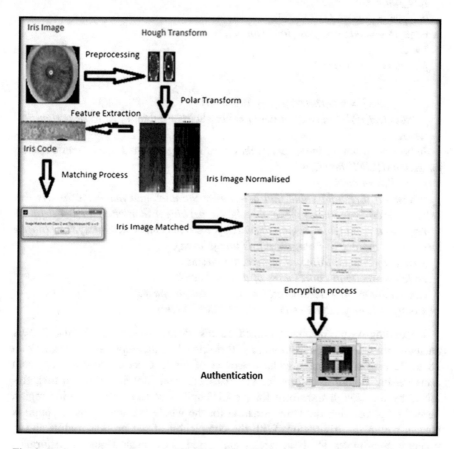

Fig. 3 Basic operations of neoteric iris subtlety

(a) Iris Segmentation/Disjuncture

A circular Hough transform is explored for detecting the iris and pupil boundaries, which involves canny edge detection to generate an edge map. Gradients are biased in the vertical direction for the outer iris/sclera boundary. Vertical and horizontal gradients are weighted equally for the inner iris/pupil boundary [1–7]. A modified version of canny edge detection MATLAB® function is implemented which allows weighting of the gradients.

The following function is developed in MATLAB for performing iris disjuncture process through Hough transform with canny edge detection:

function[x_in,y_in,radi_in,x_out,y_out,radi_out,size_reduction]=fun_iris_extrac-tion(img,fig_show)
rows = size(img,1);
size_reduction = rows/150;
re_size_img = imresize(img,1/size_reduction);
[row1,col1] = size(re_size_img);
*row_lim = row1*0.1;*
*col_lim = col1*0.1;*
edge_img = edge(re_size_img,'canny');
i = 0;
for radis = 100:-1:10
  i = i + 1;

[y0,x0,accu] = houghc(edge_img,radis,4,3)
" [y0detect,x0detect,Accumulator] = houghcircle(Imbinary,r,thresh,region)"
where
Imbinary - a binary image pixels that have value equal to 1 are interested pixels
for HOUGHLINE function.
r - radius of circle.
thresh - a threshold value that determines the minimum number of pixels that
belong to a circle in image space; threshold must bigger than or equal to 4(default).
  region - a rectangular region to search for circle centres within[x,y,w,h] (not be
larger than the image area, default is image area).
y0detect - row coordinates of detected circles.
x0detect - column coordinates of detected circles.
Accumulator - the accumulator array in Hough space.
Canny = Canny edge detection MATLAB® function

Depending on the database utilized, radius values are set, like for the CASIA database iris radius range from 90 to 150 pixels while the pupil radius ranges from 28 to 75 pixels. For making the detection of circle process more efficient and accurate, the Hough transform for the iris/sclera boundary is performed first, followed by the Hough transform for the iris/pupil boundary within the iris region instead of performing the transformation for the whole eye region, as the pupil is always within the iris region. With the completion of the process, radius and x, y centre coordinates for both circles are stored. The linear Hough transform is utilized for isolating the eyelids by first fitting a line to the upper and lower eyelids. Then, the next another horizontal line is drawn intersecting the first line at the iris edge which is closest to the pupil that allows maximum isolation of eyelid regions. An edge map is created by canny edge detection utilizing only horizontal gradient information. The function developed in MATLAB for performing the linear Hough transform specifies that if the maximum Hough space < a set threshold, no line is fitted as it corresponds to non-occluding eyelid and the lines are restricted to lie exterior to the pupil region and interior to the iris region. Linear Hough transform is considered more advantageous over its counterpart which is the parabolic version, as it has less parameter to deduce, resulting in computationally less demanding process. The results of segmentation carried out using Hough transform utilizing canny edge detection proved to be successful providing accuracy of 100%. All the images specified from standardized database (CASIA or LEI) were segmented quite efficiently with the proposed algorithm providing discrete and clear inputs for the normalization state. As normalization process needs inputs i.e. area of interest selected in the segmentation stage and if segmentation not done properly will affect the other stages and will not lead to the development of an efficient iris perception approach. Figure 4 depicts iris image and its segmentation results.

The images belonging to class 6 of the database of the proposed methodology are selected, and segmentation is performed on them through Hough transform and canny edge detection methods. The proposed methodology successfully carries out iris segmentation of all the images specified in the database, extracting the area of interest from the iris image and removing all the unwanted parts from it, resulting in an accurate input for the next stage of iris perception approach.

(b) Normalization

Normalization is performed through histogram equalization method. When compared with the other methods of performing normalization available in the literature, histogram equalization enhances the contrast of iris images by transforming the values in an intensity image so that the histogram of the output iris image approximately matches a specified histogram. This method usually increases the global contrast of iris images, especially when the usable data of the image is represented by close contrast values. By performing this adjustment, the intensities

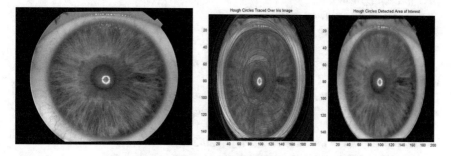

Fig. 4 Iris image segmented through Hough transform depicting areas of interest

can be better distributed on the histogram allowing areas of lower local contrast to gain a higher contrast. Histogram equalization accomplishes this task by effectively spreading out the most frequent intensity values.

To evaluate histogram equalization for iris images, let f be a given iris image represented as m_r by m_c matrix of integer pixel intensities ranging from 0 to $L - 1$, where L is the number of possible intensity values, often 256. Let p denote the normalized histogram of f and p_n = number of pixels with intensity n/total number of pixels and $n = 0, 1... L - 1$. The histogram equalized iris image g will be defined by:

$$g_{ij} = \text{floor}\left((L-1) \sum_{n=0}^{f_{ij}} p_n \right) \tag{1}$$

where floor() rounds down to the nearest integer.

This is equivalent to transforming the pixel intensities, k of f by the function:

$$T(k) = \text{floor}\left((L-1) \sum_{n=0}^{k} p_n \right) \tag{2}$$

The motivation for this transformation is the assumption of the intensities of f and g as continuous random variables X, Y on $[0, L - 1]$ with Y defined by:

$$Y = T(X) = (L-1) \int_{0}^{x} pX(x)\mathrm{d}x \tag{3}$$

where

pX probability density function of f.
T cumulative distributive function of X multiplied by $(L - 1)$.

The following function is developed in MATLAB for performing normalization process through histogram equalization for the proposed methodology:

```
function[data_stream,mask_stream,norm_ract_iris]=fun_create_pal
(img_path,fig_show)
  img_rgb = imread(img_path);
  img = rgb2gray(img_rgb);
  norm_img = fun_pre_processing(img);
  [x_in,y_in,radi_in,x_out,y_out,radi_out,size_reduction]=fun_iris_extraction
(img,fig_show);
  ract_iris = circle_to_strip([x_in*size_reductiony_in*size_reduction],radi_in*-
size_reduction,radi_out*size_reduction,norm_img);
  ract_iris_1 = ract_iris';
  resize_ract_iris = imresize(ract_iris_1,[20 200]);
  norm_ract_iris = fun_pre_processing(resize_ract_iris);
```

[data_stream,mask_stream] = fun_bin_stream(norm_ract_iris);
%imwrite(norm_ract_iris,'palette.bmp','bmp');
if fig_show == 1
figure;
colormap(gray);
subplot(1,2,1);-
imagesc((ract_iris_1));
title 'Iris Template'
colormap(gray);
subplot(1,2,2);
imagesc(norm_ract_iris); title 'Iris Template Normalized' end

When compared with other conventional iris recognition algorithms, proposed normalization process is able to perfectly reconstruct the same pattern from images with varying amounts of pupil dilation, as deformation of the iris results in small changes of its surface patterns. Even in the images where pupil is smaller, the proposed normalization process is able to rescale the iris region to achieve constant dimension. The proposed methodology generates the rectangular representation from 10,000 data points in each iris region taking into account all the rotational inconsistencies (misalignment in the horizontal (angular) direction). Rotational inconsistencies will be accounted in the matching stage. The output of the previous segmentation stage where area of interest was selected by a perfect Hough transform and canny edge detection provided the input for the normalization stage. Normalization if not perfectly done will lead to an inappropriate matching that will not build a perfect security solution which is the key requirement of this research study. Figure 5 depicts iris image with its perfect normalization performed that further becomes the input template for next stage.

(c) **Template Formation or Encoding**

Template formation or encoding in the proposed methodology is performed through convolving the normalized iris pattern with bi-orthogonal wavelets 3.5. The 2D normalized pattern is broken up into 1D signals. The angular direction is taken rather than the radial which corresponds to columns of the normalized pattern, as the maximum independence occurs in the angular direction. Transformation of the segmented iris information into a normalized iris data is done using the bi-orthogonal tap. In the proposed method rather than utilizing the traditional

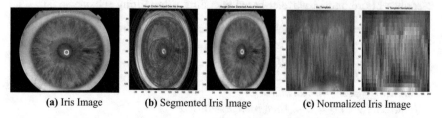

    (a) Iris Image   **(b)** Segmented Iris Image   **(c)** Normalized Iris Image

Fig. 5 Normalization of iris images

Multi-Resolution Analysis (MRA) scheme, a novel lifting technique is explored for the construction of bi-orthogonal filters. The main advantage of this scheme over the classical construction methods is that it does not rely on the Fourier transform and results in faster implementation of wavelet transform. The basic concept of lifting scheme starts with a trivial wavelet referred as the "Lazy wavelet", which has the formal properties of basic wavelet but is not capable of performing the analysis [8–10]. A new wavelet having improved properties is gradually developed by adding a new basis function which is the main inspiration behind the name of the scheme. The lifting scheme can be visualized as an extension of the Finite Impulse Response (FIR) schemes, where for any two-channel, FIR sub-band transform can be factored into a finite sequence of lifting steps making the implementation of these lifting steps faster and efficient.

Bi-orthogonal 3.5 tap is selected for encoding the iris information by adjusting the frequency content of the resulting coefficients to get a separated band structure. In the lifting scheme, the filters are designed using the lifting steps as they are completely invertible. Transformation of the data into a different and new basis is performed by the filters, where large coefficients correspond to relevant image data and small coefficients correspond to the noise. Thresholding is performed once again and referred as image de-noising. The data encoded by the wavelet is scalable and localized, making matching possible of the features at same location using various scales resulting in information of bitstream of 1 and 0 s referred as the "iris template". For performing comparison band pass Gabor pre-filtering is done to encode the information and generate the filter using Gaussian filters and then this approximation is utilized for generating wavelet coefficients that are quadrature quantized, resulting in information of bitstream of 1 and 0 s. This is performed for all the iris images, and the formulated bit pattern is referred as the "iris template" having angular resolution as 20, radial resolution as 200 and length as 8000 bits. The noisy parts of the image are located by the mask template that is formed along with the iris template.

For performing wavelet analysis in this particular study, the digital iris images are encoded using wavelets to formulate the iris template. The intensity values at known noise areas in the normalized pattern are set to the average intensity of surrounding pixels to prevent influence of noise in the output of the filtering. Phase quantization at the output of filtering is then done to four levels, where two bits of data for each phasor are produced; then, the output of phase quantization is chosen to be a grey code, so that when going from one quadrant to another, only 1 bit changes, hence minimizing the number of bits disagreeing. Algorithm below describes bi-orthogonal wavelet-based encoding using lifting scheme:

```
function [data_stream,mask_stream] = fun_bin_stream(ract_iris)
ract_iris = double(ract_iris);
els = {'p',[-0.125 0.125],0};
lsbiorInt = liftwave('bior3.5');
lsnewInt = addlift(lsbiorInt,els);
[CA,CH,CV,CD] = lwt2(ract_iris,lsnewInt);
[THR,SORH,KEEPAPP] = ddencmp('den','wv',ract_iris);
```

```
C = [CA CH; CV CD];
C1 = abs(C);
C_data = (C1 >= THR).*C;
max_val = max(C_data(:));
min_val = min(C_data(:));
C_data_norm = (C_data - min_val)/(max_val - min_val);
C_data_quant = round(C_data_norm*3);
rep = [0 0;0 1;1 0;1 1];
for i = 1:size(C_data_quant,1)
   for j = 1:size(C_data_quant,2)
      C_data_quant_new(i,2*(j-1)+1:2*j) = rep(C_data_quant(i,j)+1,:);
   end
end
C_noise = (C1 < THR).*C;
max_val = max(C_noise(:));
min_val = min(C_noise(:));
C_noise_norm = (C_noise - min_val)/(max_val - min_val);
C_noise_quant = round(C_noise_norm*3);
for i = 1:size(C_data_quant,1)
   for j = 1:size(C_data_quant,2)
      C_noise_quant_new(i,2*(j-1)+1:2*j) = rep(C_noise_quant(i,j)+1,:);
   end
end
data_stream = C_data_quant_new(:);
mask_stream = C_noise_quant_new(:);
end
```

Iris templates generated after encoding process are shown in Fig. 6. Bit-wise iris templates contain number of bits of information. The total number of bits in the template = 2* the angular resolution times the radial resolution * number of filters utilized.

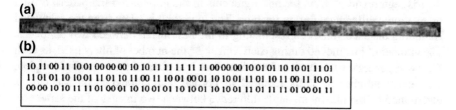

(a)

(b)

```
10 11 00 11 10 01 00 00 00 10 10 11 11 11 11 11 00 00 00 10 01 01 10 10 01 11 01
11 01 01 10 10 01 11 01 10 11 00 11 10 01 00 01 10 10 01 11 01 10 11 00 11 10 01
00 00 10 10 11 11 01 00 01 10 10 01 01 10 10 01 10 10 01 11 01 11 11 01 00 01 11
```

Fig. 6 a Iris image templates generated after encoding process, **b** code of iris template being generated

2.1 Matching Process

Matching unlike other conventional eye recognition processes is not done by taking only one single matching criterion into consideration, but in the proposed methodology, matching is performed through two parameters. Hamming distance as well as normalized correlation coefficient is utilized as metrics for recognition since bit-wise comparisons are necessary. Noise masking is incorporated by the Hamming distance algorithm so that only significant bits are used in calculating HD. The Hamming distance is calculated using only the bits generated from the true iris region, and thus, the modified HD formula is given as:

$$HD = \frac{1}{N - \sum_{k=1}^{N} X_{nk}(\mathrm{OR})Y_{nk}} \sum_{j=1}^{N} X_j(\mathrm{XOR})Y_j(\mathrm{AND})X_{n'j}(\mathrm{AND})Y_{n'j} \qquad (4)$$

where

X_j and Y_j two bit-wise templates to compare.
$X_{n'j}$ and $Y_{n'j}$ corresponding noise masks for X_j and Y_j.
N number of bits represented by each template.

Although in theory two iris templates generated from the same iris will have a Hamming distance of 0.0, in conventional methods this has not happened as normalization is not perfect and also there will be some noise that goes undetected, so some variation will be present when comparing two intra-class iris templates. But in this neoteric approach, normalization is perfectly done and also encoding done through bi-orthogonal wavelets providing a bitstream away from noises. Hence, Minimum Hamming Distance = 0 and Maximum Normalized Correlation Coefficient = 1 both are achieved. For taking into account the rotational inconsistencies, the Hamming distance of two templates is calculated and one template is shifted left and right bit-wise. Hamming distance values are calculated from successive shifts, i.e. bit-wise shifting in the horizontal direction corresponding to rotation of the original iris region by the angular resolution used [11–15]. If an angular resolution of 180 is utilized, each shift will correspond to a rotation of 2° in the iris region and corrects for misalignments in the normalized iris pattern caused by rotational differences during imaging. The lowest bit is taken from the calculated Hamming distance values corresponding to the best match between two templates. The number of bits moved during each shift = 2* the number of filters used, as each filter will generate two bits of information from one pixel of the normalized region. The actual number of shifts required to normalize rotational inconsistencies is determined by the maximum angle difference between two images of the same eye, and one shift is defined as one shift to the left, followed by one shift to the right. The shifting process performed in HD calculation is illustrated in Fig. 7.

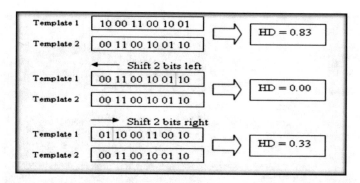

Fig. 7 Shifting process in Hamming distance calculation

The next matching parameter utilized is the normalized correlation (NC) betwixt the acquired and database representation for goodness of match which is represented as:

$$\sum_{i=1}^{n}\sum_{j=1}^{m}\frac{(P_1[i,j]-\mu_1)(P_2[i,j]-\mu_2)}{nm\sigma_1\sigma_2} \tag{5}$$

where

P_1 and P_2 images of size $n_x m$
μ_1 and σ_1 mean and standard deviation of P_1
μ_2 and σ_2 mean and standard deviation of P_2.

Normalized correlation is advantageous over standard correlation as it is able to account for local variations in image intensity that corrupts the standard correlation calculation. The proposed iris perception approach is able to achieve ideal value of matching, i.e. Maximum Normalized Correlation = 1.

Three iris images (having specular reflections) of class 6 belonging to the database of the proposed methodology are taken and undergone various successful stages of the neoteric iris perception approach. Figure 8 depicts that the selected iris image matches with the perfect class 6 achieving ideal value of matching, i.e. Minimum Hamming Distance = 0.

Main achievement of the proposed crypt-iris-based perception methodology for MANET is taking into account not one but two matching parameters, hence enhancing the proposed approach. Figure 9 illustrates that the selected iris image matches with perfect class achieving ideal value of matching, i.e. Maximum Normalized Correlation = 1 which is not achieved by any conventional iris recognition algorithms. The next section focuses on the concept of threshold cryptography, and the subsequent subsection utilizes the iris template generated by neoteric iris perception approach to produce domains of elliptic curve cryptography for further authentication. Hence, a novel "crypt-iris-based authentication approach" is developed which further enhances security of the proposed iris

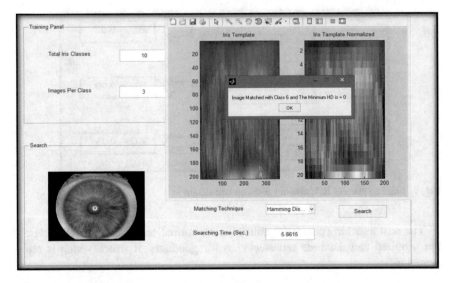

Fig. 8 Matching result with Hamming distance calculation

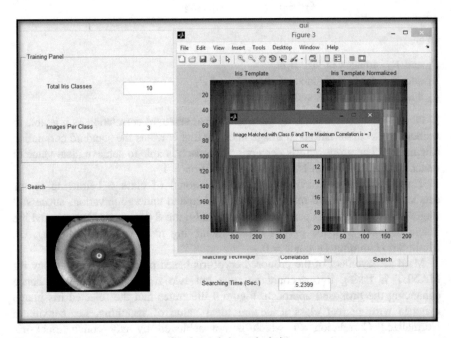

Fig. 9 Matching result with normalized correlation calculation

perception approach discussed in previous section and subsections through elliptic curve cryptography, resulting in a neoteric "crypt-iris-based perception and authentication approach", leading to an enhanced and effective security solution for

MANET utilizing traits of both biometric and cryptography which doubly secures MANET.

3 Conclusion and Future Analysis

This neoteric study has achieved successful performance parameter results, which is quite effectively depicted in the previous sections. Results achieved by "neoteric iris acclamation subtlety" are being summarized below to validate the effectiveness of the developed approach.

- **A flexible simulation environment of the iris perception approach**: it allows varying of the iris classes as well as images per class, providing effective values for various specificity and sensitivity parameters such as TPR, TNR, FPR, FNR, Precision, Accuracy, Recall and F-Measure.
- **Time for training:** various iris classes is **not very high** even with increase in the number of iris classes and images per class.
- **Approximately very accurate values** of TPR = nearly 100%, TNR = nearly 100%, FPR = nearly 0%, Accuracy = 100%, Recall = 100% and F-Measure = Nearly 100% are achieved by the neoteric iris perception approach for MANET.
- When compared with Masek [16] work on iris recognition which achieved **FNR and FPR** (with different classes per samples) as **4.580 and 2.494** on LEI database and **5.181 and 7.599** on CASIA database, the **proposed methodology** serves as a neoteric approach achieving required values of **FNR = 0 and FPR = 0.012346** (many parameters included in the proposed methodology are not being specified by any of the conventional approaches) leading to enhanced security solution for MANET.
- In the **genetic stowed approach, all the parameters** such as network length, network width, nodes transmission range, nodes traffic, speed variation factor and angle variation factor in the network configuration panel as well as number of nodes, simulation time and averaging loop value in the simulation panel can be **flexibly varied**, providing effective solutions and maintaining QOS which is the most critical requirement for MANET.
- **Low packet drop rate** of the proposed methodology compared with conventional shortest path selection method (Dijkstra's) having a high packet drop rate with different comparable parameters, namely node transmission range, node speed, node data rate and node traffic. Even with increase in number of nodes, simulation time and various other parameters, results are not deteriorated.
- **High packet delivery ratio** of the proposed methodology compared with conventional shortest path selection method (Dijkstra's approach) having a low

packet delivery ratio with different comparable parameters, namely node transmission range, node speed, node data rate and node traffic. Even with increase in the number of nodes, simulation time and other parameters, results are not deteriorated.

- **Lower end-to-end delay** of the proposed methodology compared with conventional shortest path selection method (Dijkstra's approach) having a higher end-to-end delay with different comparable parameters, namely node transmission range, node speed, node data rate and node traffic. Even with increase in the number of nodes, simulation time and various other parameters, results are not deteriorated.

- **Lower hop counts** required for transferring packets from source to destination of the proposed method compared with conventional shortest path method having larger number of hop counts with different comparable parameters, namely node transmission range, node speed, node data rate and node traffic. Even with increase in the number of nodes, simulation time and various other parameters, results are not deteriorated.

References

1. J. Daugmann, *Proceedings of International Conference on Image Processing* (2002)
2. P. Khaw, in *Iris Recognition Technology for Improved Authentication*. SANS Security Essential (GSEC) Practical Assignment, Version 1.3, SANS Institute, 5–8 (2002)
3. J. Daugmann, High confidence visual recognition of persons by a test of statistical independence. IEEE Trans. 1 Pattern Anal. Mach. Intell. **15**(11) (1993)
4. J. Daugmann, *Biometrics: Personal Identification in Networked Society* (Kluwer Academic Publishers, 1999)
5. J. Daugmann, Iris recognition. Am. Sci. **89**, 326–333 (2001)
6. J. Daugmann, in *How Iris Recognition Works*. IEEE Transactions on Circuits and Systems for Video Technology (CSVT), vol. 14, no. 1, pp. 21–30 (2004)
7. J. Daugmann, in *Probing the Uniqueness and Randomness of Iris Codes: Results from 200 Billion Iris Pair Comparisons*. Proceedings of the IEEE, vol. 94, pp. 1927–1935 (2006)
8. R.P. Wildes, in *Iris Recognition: An Emerging Biometric Technology*. Proceedings of the IEEE, vol. 85, no. 9, pp. 1348–1363 (1997)
9. R.P. Wildes, J.C. Asmuth, C. Hsu, R.J. Kolczynski, J.R. Matey, S.E. McBride, Automated, non invasive iris recognition system and method. U.S Patent (5,572,596) (1996)
10. M. Misiti, Y. Misiti, G. Oppenheim, J. Poggi, *Wavelet Toolbox 4 User's Guide* (1997–2009)
11. S. Zafar, M.K. Soni, M.M.S. Beg, in *An Optimized Genetic Stowed Biometric Approach to Potent QOS in MANET*. Proceedings of the 2015 International Conference on Soft Computing and Software Engineering (SCSE'15), Procedia Computer Science, vol. 62, pp. 410–418 (Elsevier, 2015a)
12. S. Zafar, M.K. Soni, Biometric stationed authentication protocol (BSAP) inculcating meta-heuristic genetic algorithm. I. J. Mod. Edu. Comput. Sci. 28–35 (2014a)
13. S. Zafar, M.K. Soni, A novel crypt-biometric perception algorithm to protract security in MANET. Genetic algorithm. I. J. Comput. Netw. Inf. Secur. 64–71 (2014b)

14. S. Zafar, M.K. Soni, in *Secure Routing in MANET Through Crypt-Biometric Technique*. Proceedings of the 3rd International Conference on Frontiers of Intelligent Computing: Theory and Applications (FICTA), pp. 713–720 (2014c)
15. S. Zafar, M.K. Soni, M.M.S. Beg, in *QOS Optimization in Networks through Meta-heuristic Quartered Genetic Approach*. ICSCTI (IEEE 2015b)
16. L. Masek, *Recognition of Human Eye Iris Patterns for Biometric Identification* (University of West California, 2003)

Performance Comparison of Hampel and Median Filters in Removing Deep Brain Stimulation Artifact

Manjeet Dagar, Nirbhay Mishra, Asha Rani, Shivangi Agarwal
and Jyoti Yadav

Abstract The principal purpose of this paper is to incorporate digital filters in the preprocessing of electroencephalogram (EEG) signal to remove deep brain stimulation (DBS) artifact. DBS is used in the treatment of Parkinson's disease (PD). During the monitoring of EEG, various stimulation artifacts may overlap with EEG signal. Therefore, a filter is required, which can effectively eliminate these artifacts with the least distorting EEG signal. In the present work, performance comparison of Hampel and median filters is carried out to eliminate these artifacts. The effectiveness of these filters is tested on different types of signals: sinusoidal, synthetic EEG signals. Further, these filters are tested on real EEG signals corrupted with DBS noise. These signals are acquired from patients under the treatment for PD. The performance comparison of filters is evaluated on the basis of signal-to-noise ratio (SNR), SNR improvement (SNRI), mean square error (MSE), and signal distortion. The results reveal that Hampel filter removes the noise more efficiently as compared to median filter.

Keywords Hampel filter · Median filter · EEG · Parkinson's disease
Deep brain stimulation artifact

M. Dagar (✉) · N. Mishra · A. Rani · S. Agarwal · J. Yadav
Netaji Subhas Institute of Technology, University of Delhi, Delhi, India
e-mail: manjeetdagar77@gmail.com

N. Mishra
e-mail: nirbhay2005@gmail.com

S. Agarwal
e-mail: agarwal.shivangi@gmail.com

J. Yadav
e-mail: bmjyoti@gmail.com

© Springer Nature Singapore Pte Ltd. 2018
B. Panda et al. (eds.), *Innovations in Computational Intelligence*, Studies in
Computational Intelligence 713, https://doi.org/10.1007/978-981-10-4555-4_2

1　Introduction

PD is a disease of progressive Neuro-decision of the central nervous system (CNS) that mainly has an impact on the motor system. In the initial stages of the disorder, the symptoms, which are most common, are movement-related which include slowness of movement, stringency, shaking and difficulty while walking, and amble or balance-related problems. In some cases, behavioral problems may also arise, with dementia mostly occurring in the advanced course of the PD, and the most common psychiatric symptom is depression. Sleep, sensory, and emotional problems are other symptoms that may be observed in PD [1].

Presently, the DBS is used only for patients whose symptoms cannot be adequately controlled with medications. There is a set of nuclei just under the ganglia nuclei which are quite deep in the brain [2]. Then, the surgeon will put a small hole either on one side or both sides; tunnel these electrodes on both sides with each side containing four electrodes which are able to stimulate that nucleus as the electricity itself enhances the communication. The leads are implanted under the neck and a generator is implanted in the chest as shown in Fig. 1. Through this generator, a neurologist or a programmer can initiate the process. It will deliver a certain amount of current at certain frequencies and manipulate those four electrodes in an optimal way for treatment.

Effect of DBS on the cortex is mainly studied through imaging modalities. In order to get information about movement-related activities, temporal information needs to be monitored with the help of EEG. While monitoring EEG signals, electrical noise due to DBS is overlapped with EEG spectrum. Even if the DBS is

Fig. 1 Implanted DBS simulator [3]

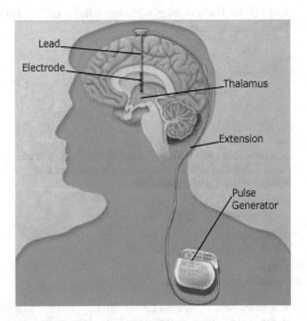

off, there is some aftereffects of DBS simulator. It is difficult to remove frequency of DBS artifact from the EEG signal as both are having same frequency range; it is suggested that a digital filter should be selected carefully to enhance the quality of the signal. Ronald et al. [4] discussed the class of generalized median filters: recursive median filter, weighted median filter, recursive median filter, and Hampel filter. The generalized Hampel filter has additional parameter threshold (t) which gives it more degrees of freedom. Allen et al. [5] described the use of Hampel filter for off-line frequency-domain filtering for the removing outliers. A time-domain signal is converted into the frequency domain with the fast Fourier transform (FFT) which detects outliers, replaces those outliers with median values, and transforms the cleaned spectra into the time domain using the inverse FFT. One of the best feature of Hampel filter is that it does not require prior knowledge about the frequency content of signal which makes it suitable for application such as removing DBS noise from EEG signal [6]. Rincon et al. [7] used Hampel filter for EEG preprocessing for removal of different types of artifacts present in EEG. Zivanovic et al. [8] utilized the Hampel filter for removing power line interference (PLI) and baseline wandering (BW) of ECG and EMG signals. A comparative study was performed using different methods: adaptive filter, Hampel filtering, and subtraction method. The Hampel filter provides the best performance among these methods in the whole analysis range. Hampel filter can mitigate the impact of PLI components on the ECG signal by minimizing the sharp peaks. Beuter et al. [9] demonstrated the application of the median filter to study the effect of deep brain stimulation (DBS) on amplitude and frequency characteristics of tremor in patients suffering from Parkinson's disease.

The present work focuses on the preprocessing of EEG signal to filter out DBS artifact induced due to electrical noises with least distortion in the signal. An efficient filter for data cleaning is one which satisfies two important criteria; firstly, it can remove the outliers, i.e., the abnormal or extreme values in our observation, with more appropriate data values that are consistent with the overall sequence of our observations, and secondly, it must not alter values which are appropriate, i.e., signals which are in our required specified range. The required digital filter should function such that no changes or very few changes occur, in the nominal data sequence [1].

In the present work Hampel, the filter is applied to the signals contaminated with outlier, and the results are compared with the median filter, which is a standard filter for data cleaning. From the literature, it is suggested that Hampel filtering is superior to simple standard median filter, in the sense that it preserves the low-level signal. Moreover, Hampel filter causes minimal phase distortion to the signal, which makes it ideal filter for removing outliers.

2   Hampel Filter

Hampel filter is a nonlinear filter of data cleaning that searches abnormal local data in a time sequence, and these values are substituted by more reasonable alternative values once found; the algorithm uses the median absolute deviation (MAD) in temporal series, i.e., it uses the median of specified neighboring values as reference value and MAD scale estimator as a measurement of the deviation from the normal value. In other words, the central value in the data is considered as an outlier and replaced if it is more than T times the moving data window's MAD scale estimate from its median:

$$y_k = \begin{cases} x_k & \text{if } |x_k - m_k| \leq tS_k \\ m_k & \text{otherwise} \end{cases} \tag{1}$$

Here,

m_k is the median of w_k
S_k is the MAD scale estimate, defined as:

$$S_k = 1.4826 \times \text{median} j \in [-K, K] \{|x_k - j - m_k|\} \tag{2}$$

Note that m_k is the output of the standard median filter, hence Hampel filter acts as a standard median filter at $t = 0$. Here K is the window length, and T is the threshold parameter.

2.1   Performance Evaluation Measures

For the comparison of the performance of the Hampel and median filters, the quantitative analysis is performed using different parameters, i.e., MSE, SNR, SNRI, and signal distortion using SINAD [2].

MSE: It defines the difference among the filtered signal and reference signal, and the mathematical relation is given below:

$$\text{MSE} = \frac{1}{K} \sum_{n=1}^{K} (\text{signal}_{\text{ref}} - \text{signal}_{\text{filtered}})^2 \tag{3}$$

The low value of *MSE* indicates that the quality of the signal is good and vice versa.

SNR: Signal-to-noise ratio is an important parameter for artifact analysis. It is the ratio of the power of the original signal to the power of noise signal.

$$SNR = \frac{Signal_{power}}{Noise_{power}} \qquad (4)$$

The high value of SNR indicates that the quality of signal is good and vice versa.

Signal Distortion: It is calculated with the help of SINAD function. SINAD is defined as the signal-to-noise-and-distortion ratio. Signal Distortion is calculated when the SNR is subtracted from SINAD. Mathematically, it is computed as:

$$Signal\ Distortion = SINAD - SNR \qquad (5)$$

SNRI: This parameter indicates the improvement of output SNR over input SNR. SNRI is calculated in the equation below. A higher value of SNRI shows enhancement in the signal quality, while negative SNRI denotes that SNR input is deprecated because of denoising mechanism.

$$SNRI = SNR_{out} - SNR_{in} \qquad (6)$$

For the verification of the viability of Hampel filter in preprocessing of signal, it is carried out on different signals. The first data type uses a sinusoidal signal added with noise and checks for the optimal value of the function for different values of threshold parameter (T) and half window length (DX) for the Hampel filter. The second data type uses a synthetic EEG signal with noise added to it. The performance comparison of Hampel filter is done with the median filter. The last data in our observations are real EEG contaminated with (DBS) artifact obtained from different subjects who are using DBS for the treatment of PD. In next section, the performance of both the filters is evaluated.

3 Results and Discussion

3.1 Hampel Filter Design for Sinusoid

For the application of Hampel filter, selection of T and DX is the primary concern. The width of the window is given by $W = 2DX + 1$ [2]. For higher values of DX, window length W enhances which results in a reduction in SNRI. When T is increased for fixed window length, MSE increases. Hence, an optimum selection of DX and T is required to minimize MSE and improve the SNRI. A thorough analysis is performed to produce the optimum values of DX and T. The value of DX is varied in the wide range, while T is varied from 0.01 to 0.3, as beyond these values the error multiplies unusually. The results are undesirable for all the combinations of T and DX. The observations show that SNRI decreases with rising in T, whereas the MSE follows a reverse trend.

Firstly, a sinusoidal signal is used to perform the experiment. Some outliers are added into the sinusoidal signal. This signal is now treated as signal contaminated

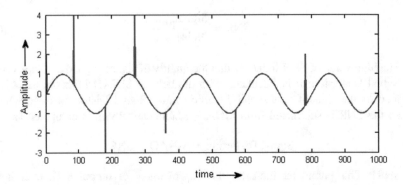

Fig. 2 Sinusoidal wave with artifacts

Table 1 Comparative results of Hampel and median filters in case of sinusoidal signal

DX	T	Hampel filter			Median filter		
		SNRI	MSE	Distortion	SNRI	MSE	Distortion
5	0.05	39.19	0.0438	−1.2204			
10	0.1	32.80	0.0437	−2.1968	39.23	0.550	−0.1947
15	0.3	27.14	0.0439	−3.2379			

with outliers. This signal containing artifacts is now filtered using Hampel filter and median filters. The primary objective of the filtration process is to minimize the effect of noise and artifacts. The original signal and noisy signal at the SNR of 10 dB are shown in Figs. 1 and 2, respectively. Here, Hampel filter is applied to a noisy sinusoidal signal. The filter parameters are adjusted to get as much lower values of MSE and signal distortion and as much greater value of SNR as possible [10]. The table shows the performance of Hampel filter for different values of DX and T. The performance of Hampel filter is compared with a median filter on the basis of above-mentioned parameters. It is revealed from Table 1 that MSE and signal contortion are notably minimized by using Hampel filter as compared to median filter. The reason for this reduction is that Hampel filter has more flexibility with the parameters as compared to the median filter. It is observed that the results obtained are highly favorable, and hence it is clear that Hampel filter can further be applied for the preprocessing of EEG signals (Fig. 3).

3.2 Hampel Filter Implementation on Synthetic EEG

Performance comparison of the filters has been verified by using an EEG signal which is produced synthetically, and outliers are incorporated in the signal. In this paper, the synthetically produced EEG signal is created using 7 segments [2].

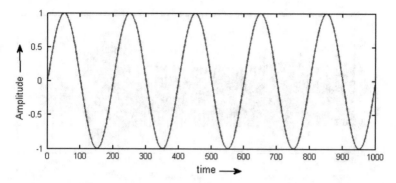

Fig. 3 Filtered sinusoidal signal

The segments differ in frequency and amplitude. The sampling frequency is 250 Hz and is simulated for the same time span.

Details are given below [1].

Part 1: $4\cos(4\pi T) + 4\cos(8\pi T)$
Part 2: $1.5\cos(\pi T) + 4\cos(4\pi T)$
Part 3: $0.5\cos(2\pi T) + 1.5\cos(5\pi T) + 4\cos(8\pi T)$
Part 4: $0.7\cos(2\pi T) + 2.1\cos(5\pi T) + 5.6\cos(8\pi T)$
Part 5: $1.5\cos(3\pi T) + 1.9\cos(5\pi T) + 4.7\cos(5\pi T)$
Part 6: $4.5\cos(3\pi T) + 9.8\cos(9\pi T)$
Part 7: $1.7\cos(\pi T) + \cos(4\pi T) + 5\cos(6\pi T)$

All the individually specified signals are used to compose a synthetic EEG signal. The segments 1 and 2 are different only in terms of frequency, signals 3 and 4 are different only in amplitude, and the remaining signals differ in amplitude as well as frequency. Hence, the synthesized signal contains all the possible states [1].

The process of implementation of the Hampel filter on the synthetic EEG signal involves two steps. In the first step, sine wave as synthetic artifacts is added to the EEG signal.

Firstly, the contaminated EEG signal is filtered by Hampel filter, and the optimum values of parameters are found out by adjusting the values of DX and T. Then, the results are compared with median filter by the use of MSE, SNR, and signal distortion values. The table shows the values of different parameters for both the filters. It is evident from the observations that the Hampel filters perform better than median filter as it has lower values of MSE and higher values of SNRI (Figs. 4, 5, 6, and 7; Table 2).

In the next step, synthetic EEG signal is contaminated with sine artifacts. The filtered output is compared with a median filter on the basis of performance parameters. The observations show that the performance of Hampel filter is better than that of the median filter as it has better SNR and much lesser MSE. The one factor that helps in better performance of Hampel filter is that the flexibility of

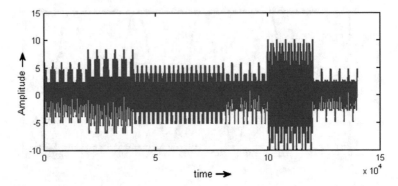

Fig. 4 Synthetic EEG signal

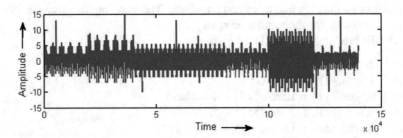

Fig. 5 Synthetic EEG contaminated with outliers

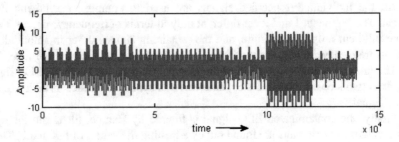

Fig. 6 Filtered EEG signal with Hampel filter

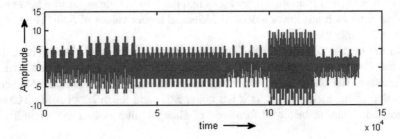

Fig. 7 Filtered EEG signal with Median filter

Table 2 Comparative results of Hampel and Median filters in case of synthetic EEG signal

DX	T	Hampel filter			Median filter		
		Signal			Signal		
		SNRI	MSE	Distortion	SNRI	MSE	Distortion
10	0.3	0.0185	0.0107	−1.269e−06			
15	0.3	0.0167	0.0107	−4.500e−06	0.0015	13.22	−8.13e−07
20	0.4	0.0136	0.0106	−1.78e−05			
30	2.0	0.0193	0.0109	−8.11e−07			

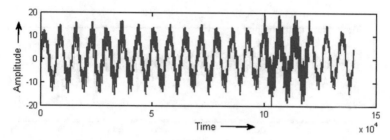

Fig. 8 Filtered synthetic EEG signal with sine artifacts

Table 3 Comparative result of synthetic EEG signal with sine artifacts

DX	T	Hampel filter			Median filter		
		MSE	SNRI	Signal distortion	MSE	SNRI	Signal distortion
30	0.025	13.206	0.0209	−0.0010			
35	0.03	13.207	0.0325	−0.0010	50.00	2.2307e−06	−0.0010
40	0.07	13.206	0.0475	−0.0011			
50	0.2	13.206	0.0886	−0.0010			
60	0.3	13.208	0.1439	−0.0010			

altering the values of window length and threshold value according to the requirement which is not the case with a median filter (Fig. 8; Table 3).

3.3 Hampel Filter Implementation for Removal of DBS Artifact

The final step involves the implementation of Hampel filter for filtering of real EEG signal with DBS artifact. The DBS signals which are applied to the patients for the treatment of PD are taken for the observation. The EEG signals are acquired from different patients [4]. Each EEG signal consists of two parts. First is a file called .rit

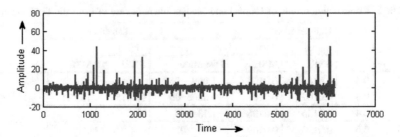

Fig. 9 Noisy EEG signal when DBS off and medication off

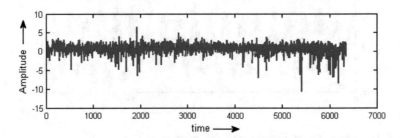

Fig. 10 Filtered EEG signal when DBS off and medication off

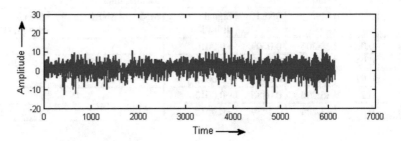

Fig. 11 Noisy EEG signal when DBS on and medication on

file with the condition "DBS off and medication off". Here the patient is not subjected to any medication or DBS signal for a minimum period of 12 h and the subject stimulator (implanted). The second file is .let file with condition "dbs on and medication on". In this state, the subject or patient had intake 150% of their morning dose of L-dopa and was receiving effective stimulation of the STN [4].

The two data types from different subjects were filtered using both Hampel filter and median filter. Here also the performance of Hampel filter is considerably better than median filter (Figs. 9, 10, 11, and 12; Table 4).

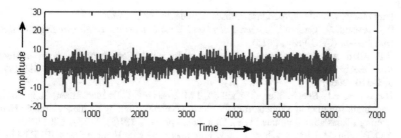

Fig. 12 Filtered EEG when DBS on and medication on

Table 4 Comparative result for removal of DBS artifact

Hampel filter						Median filter	
	File type	DX	T	SSNR	SNRI	SSNR	SNRI
P 1	DBS off and medication off	20	1	0.9873	4.8181	0.9870	0.0194
	DBS on and medication on	16	4	0.9996	0.2600	0.9906	0.0034
P 2	DBS off and medication off	20	38	0.9558	−3.31e−05	0.2877	3.8862
	DBS on and medication on	20	13	0.9614	0.2688	0.7432	−0.3641
P 3	DBS off and medication off	15	3	0.9959	0.1734	0.9704	−0.0215
	DBS on and medication on	15	5	0.9997	−0.0034	0.9830	−0.0012

4   Conclusion

In the present paper, a comparative analysis of Hampel filter and median filter is implemented in the preprocessing of real EEG signal with DBS artifact taken from different subjects. The Hampel filter is first implemented using MATLAB, and the performance of the filter is analyzed using SNR, SNRI, MSE, and signal distortion, and the results are then compared with a median filter. It is observed that Hampel filter shows superior performance as compared to median filter as it gives lesser MSE value and signal distortion value with greater SNRI. This is also due to the fact that Hampel filter has window length and threshold variables whose values can be manipulated to get optimum values of filter parameters. Thus, the application of Hampel filter may help in better understanding of the technique of action of this significant therapeutic intervention in the case of PD.

References

1. A. Beuter, M.S. Titcombe, F. Richer, C. Gross, D. Guehl, Effect of deep brain stimulation on amplitude and frequency characteristics of rest tremor in Parkinson's disease. Thalamus Relat. Syst. **1**, 203–211 (2001)
2. S. Agarwal, A. Rani, V. Singh, A.P. Mittal, Performance evaluation and implementation of FPGA based SGSF in smart diagnosti applications. J. Med. Syst. **40**, 63. doi:10.1007/s10916-015-0404-2 (2016)

3. http://www.webmd.com/parkinsons-disease/guide/dbs-parkinsons
4. P.K. Ronald, M. Gabbouj, *Median Filters and Some Extensions. Nonlinear Digital Filtering with Python* (CRC Press, 2015)
5. D.P. Allen, A frequency domain Hampel filter for blind rejection of sinusoidal interference from electromyograms (Translation Journals Style). J. Neurosci. Methods **155**(177(2)), 303–310 (2009)
6. D.P. Allen, E.L. Stegemöller, C. Zadikoff, J.M. Rosenow, C.D. Mackinnon, Suppression of deep brain stimulation artifacts from the electroencephalogram by frequency-domain Hampel filtering (Translation Journals Style). Clin. Neurophysiol. **121**(8), 1127–1132 (2010)
7. A.Q. Rincón, M. Risk, S. Liberczuk, EEG preprocessing with Hampel filters. IEEE Latin Am. Trans. Córdoba, Argentina **2012**(89), 13–15 (2012)
8. M. Zivanovic, M. González-Izal, Simultaneous powerline interference and baseline wander removal from ECG and EMG signals by sinusoidal modelling. Med. Eng. Phys. **35**, 1431–1441 (2013)
9. A. Beuter, Effect of deep brain stimulation on amplitude and frequency characteristics of rest tremor in Parkinson's disease. Thalamus Relat. Syst. **1**(3), 203–211 (2001)
10. H.J. Lee, W.W. Lee, S.K. Kim, H. Park, H.S. Jeon, H.B. Kim, B.S. Jeon, K.S. Park, Tremor frequency characteristics in Parkinson's disease under resting-state and stress-state conditions. J. Neurol. Sci. (2016)

Review on Text Recognition in Natural Scene Images

Deepti Kaushik and Vivek Singh Verma

Abstract Text recognition is a field to detect text from digital images and produces meaningful information for various applications such as navigation, object recognition, hoardings, image retrieval. This paper presents a detailed review of various existing schemes performing the text recognition operation in natural scene images. Though researchers have done phenomenal work in this area but still there is scope to ample the prick. Keeping this in mind, a comparative and quantitative analysis of various existing schemes are presented. This review work is mainly focused on connected component and region-based methods. Various advantages and limitations are also discussed in this paper. The main aim of this paper is to present literature review for researchers to serve the purpose of good reference in the areas of scene text detection and recognition.

Keywords Text recognition · Region-based and connected component-based methods · Natural scene images · Text detection · Complex background

1 Introduction

Text recognition serves as one of the important areas of pattern recognition and artificial intelligence. Text detection and recognition in natural images which are captured by camera is an important problem since text available in an image provides the useful information. Text present in images serves as an important source of information about the location from where they were clicked. Text recognition is the procedure to categorize the input character in accordance to the predefined character class. Text can in the form of scanned handwritten document or typed text

D. Kaushik (✉) · V.S. Verma
Ajay Kumar Garg Engineering College, Ghaziabad 201009, India
e-mail: myselfdeepti22@gmail.com

V.S. Verma
e-mail: viveksv10@gmail.com

© Springer Nature Singapore Pte Ltd. 2018   29
B. Panda et al. (eds.), *Innovations in Computational Intelligence*, Studies in
Computational Intelligence 713, https://doi.org/10.1007/978-981-10-4555-4_3

or a union of both. The text recognition system helps to establish a good communication between human and a computer.

In the past few years, the focus on research work is increased to detect and recognize text from natural scene images in order to attain good truthfulness. It is basically possible as its utility in various applications, such as strengthening automobile driving, navigation, and recognize objects. Moreover, it strengthens the ability of image recovery models to recover important information as it yields explanation to the matter of images.

Various approaches [1–5] have been developed so far for text detection, but still it holds to be as a interesting and unexplained issue. The reason is that the text present in scene images gets affected by difficult background, appearance dissimilarity, complex colors, different font size used in same word, uncommon font styles, inconsistent skew, various effects, and occlusion as illustrated in Fig. 1 which shows a broad range of images comprising different texts with variations discussed. Hence, we can see a large room for improvement is left in this area. This paper presents the detailed literature review on text recognition in natural scene images. The main motive to write this paper is to understand the existing schemes proposed by researchers in this field.

The paper is organized as: Sect. 2 depicts the brief description of text recognition system. Section 3 provides a detailed review of methods to detect and localize natural scene text. Section 4 gives the description about datasets and performances of various algorithms. Concluding remarks are given in Sect. 5.

Fig. 1 Text detection from images containing texts in different orientations

2 Text Recognition System

A text recognition system receives an image as an input containing some text information. The output produced by the system must be in computer readable form. The modules of the text recognition system are as following: (1) preprocessing module, (2) text recognition module, and (3) post-processing module. The overall architecture is shown in Fig. 2.

2.1 Preprocessing Module

Generally, the optical scanner is used to scan the paper document and converts it to the picture form. A picture is the collection of picture elements which are called as pixels. Up to this stage, our data is in the image form and this image is further evaluated in order to retrieve the relevant information. We execute some operations on the input image as to improve its quality and the operations are listed below:

1. Noise removal
   One of the most prime processes is noise removal. As a result, the quality of the image is enhanced and it will improve recognition process to better recognize the text in images. And it helps to yield the more exact output at the end of processing.

Fig. 2 Architecture of text recognition

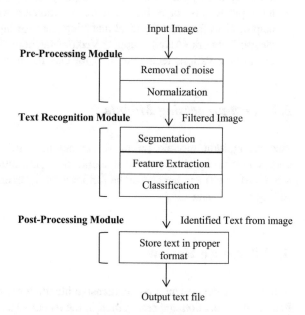

2. Normalization

   For text recognition, normalization serves as the basic preprocessing operation. In order to obtain characters of consistent size, slant, and rotation, normalization operation is executed.

2.2 Text Recognition Module

The output image of preprocessing model serves as input to this module for text recognition and gives output data in the format understandable by computer. This module uses the following techniques:

1. Segmentation

   The segmentation serves as a key process in text recognition module. This operation is executed to make a gap between the characters of an image.

2. Feature Extraction

   In order to retrieve the relevant data from the raw data, feature extraction process is applied. The relevant data means that the characters can be illustrated exactly. Various classes are made to store the different features of a character.

3. Classification

   The process of classification is done to identify each character, and the correct character class is assigned to it, in order to convert texts in images in the form understandable by the computer. Extracted features of text image are used for classification by this process, i.e., output yield by feature extraction process is the input to this stage. Input feature is compared with stored pattern with the help of classifiers and the ideal matching class for input is discovered. Artificial Neural Network (ANN), Support Vector Matching (SVM), Template Matching etc. are the various technique used for classification.

2.3 Post-processing Module

Text recognition module produces the output which is in the form text data understandable by computer, so it should be stored into some proper format (i.e., MS Word or txt) and hence, can be further used for operations such as searching or editing in that data.

3 Literature Review

In this paper, we present a comprehensive literature review on research work in the field of text detection and recognition in the recent years, mainly from the aspect of representation. This survey is dedicated for: (1) introduction of the recent works and abridge latest developments, (2) comparison of various methods and focus on up-to-date algorithms, and (3) evaluation of improvement and prediction of further research.

Various methods have been proposed by the researchers in last two decades, for identification of texts in images and videos. These methods are mainly classified into two categories given below:

- Region-based methods.
- Component-based methods.

Region-based methods are used to discriminate text areas from non-text areas in an image, local intensities, filter responses, and wavelet coefficients [6–9]. All locations and scales need to be scanned which makes these methods computationally more expensive. Also, these methods are sensitive to scale change and rotation and can handle horizontal texts best.

To localize the text in color images, a method was proposed by Zhong et al. [7]. Horizontal spatial variance was used for localization of texts approximately and then to find text within the localized regions color segmentation was implemented. Later, Li et al. [10] presented a text detection method for detecting and tracking text in videos. In this method, mean of wavelet coefficients is used for decomposition of images.

SVM classifier was trained by Kim et al. [8] to classify each pixel with the help of crude pixel intensity taken as a native feature. This technique yields great identification results in simple images (Fig. 3), but it does not works efficiently when applied complex natural images.

For multilingual texts (Chinese and English), Lyu et al. [11] suggested a raw-to-fine wide-spectrum hunt technique. To separate the non-text areas from text areas, varying contrast and sharp edge of texts are used. Additionally, this scheme yields a binary approach for segmentation of identified text regions.

To do the text recognition process rapidly, Chen et al. [6] proposed a rapid text recognition system. Cascade Adaboost [12] classifier is the detector which is utilized to train each weak classifier with the help of some attributes listed below.

- Horizontal difference,
- Gradient histogram,
- Mean strength,
- Intensity variance,
- Vertical difference.

The detection competency of this technique is considerably better than other algorithms [see for example, 13–15], but the recognition precision on real-world images is partial. To detect particular words from natural images, a method is proposed by Wang et al. [16]. Initially, single characters are extracted by sliding window. Afterward, most probable combinations are recorded based on structural relationships of characters. From the list of output result, the most alike combinations are selected. This algorithm is only capable of identifying words in the

Fig. 3 Snapshot of video dataset for text detection proposed by Kim et al. [8] showing **a,** **b** original frames and **c, d** extracted texts

specified list, and it is not able to identify words which are not in the list. It is not possible that the list contains all probable words for every image. Therefore, practically this application is less efficient.

Schemes based on connected component initially extract candidate components using different means such as color clustering or extreme region extraction, after which separation of non-text components is done utilizing automatically trained classifiers or manually designed rules [see for example, 17–20]. As the number of components to be processed is relatively less, this makes these methods more effective. These methods are also not sensitive to font variation, rotation, and scale change. Recently, these techniques have become dominant in the area of image text recognition.

A new image operator has been offered by Epshtein et al. [17] known as stroke width transform (SWT) which utilizes the characteristics that letters have almost uniform stroke width. This algorithm gives an efficient way to identify text area of various dimensions and orientations from natural images, shown in Fig. 4. This technique is applied for horizontal texts only and incorporates many manually framed directions and framework.

Fig. 4 Image dataset used in [17]

A method using the concept of maximally stable extremal regions (MSER) has been proposed by Neumann et al. [21] for text detection. Utilizing the trained classifiers, invalid candidates are excluded and extraction of MSER regions as candidates is performed from images (see Fig. 5). Afterward, the retrieved data is assembled in groups of lines by applying the sequence of predefined rules. It is not

(a)   (b)   (c)

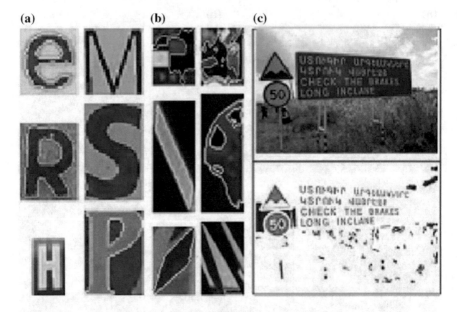

Fig. 5 Dataset showing text detection proposed by Neumann et al. [21]

suitable for higher inclination angles and can only works using horizontal or partially horizontal text.

The method proposed by Yi et al. [19] is capable of detecting tilted texts in
images. According to pixel distribution in color space, the image is first divided into
different regions, after that different regions get merged based on different features
such as relative size of regions, color similarity, and spatial distance. The
non-required components are rejected using predefined guidelines. It becomes difficult to be applied to large-scale complex image as this technique depends on a lot of
manually framed filters and parameters. To identify text which is multi-oriented,
Shivakumara et al. [22] also presents a new technique. The Fourier–Laplace space is
utilized to extract candidate regions and skeletonization is done for dividing areas
into different segments. Though these segments usually resemble to text areas rather
than characters. It does not identify characters or words directly; hence, it cannot be
quantitatively compared with other methods. Based on SWT [17], a new algorithm
that can identify texts of random alignments in natural images has been proposed by
Yao et al. [18], shown in Fig. 6. This algorithm utilizes two-step sorting method
intended for recording the basic features of text in images.

Using SWT, a new operator is proposed by Huang et al. [20] called as Stroke
Feature Transform (SFT). The SFT uses color consistency and limits association of
edge points which gives improved text extraction outcomes. For horizontal texts,
the performance of this scheme is significantly better than other methods.

A new framework which integrates convolutional neural networks (CNN) and
maximally stable extremal regions (MSER) has been proposed by Huang et al. [23]

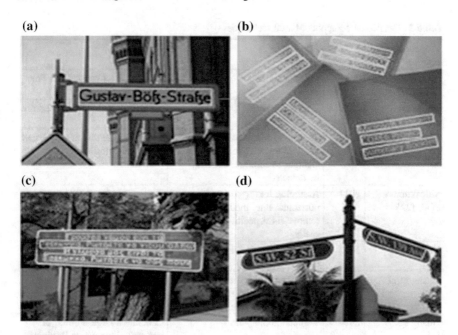

Fig. 6 Examples showing text detection proposed by Yao et al. [18]

Table 1 Region-based techniques

Proposed schemes	Year	Advantages	Limitations
Zhong et al. [7]	1995	Algorithm for locating characters is fast Algorithm is robust to variation in font, color, and size of the text	Incapable to locate text which is not well separated from the background
Li et al. [10]	2000	Can detect graphical text and scene text with different font sizes	Detects text which is horizontally aligned
Kim et al. [8]	2003	Can extract texts from complex backgrounds	Fails to detect very small text or text with a low contrast
Lyu et al. [11]	2005	Robust to background complexities and text appearance	Non-horizontally aligned text cannot be localized
Chen et al. [6]	2004	Detects text easily	Detects text which is horizontally aligned
Wang et al. [16]	2010	Detects text easily	Detects text which is horizontally aligned

for text identification in images. Firstly, text candidates are mined by using the MSER, and then, CNN-based classifier correctly identifies the text candidates and separates the connections of multiple characters. This algorithm performs significantly better than other methods. Tables 1 and 2 illustrate the strength and weakness of text detection methods discussed so far.

Table 2 Connected component-based techniques

Proposed schemes	Year	Advantages	Limitations
Epshtein et al. [17]	2010	Reasonably fast	Cannot detect curved text lines
Neumann et al. [21]	2010	Reasonably fast	Sometimes individual letters are not detected
Yi et al. [19]	2011	Text in arbitrary colors, sizes, orientations, locations, and changes in illuminations can be detected	Fails to locate text of very small size, overexposure
Shivakumara et al. [22]	2011	Assuming text appears on horizontal line, method can remove false positives more easily	Difficult to detect text from complex background
Yao et al. [18]	2012	Texts of various orientations can be detected	Rely on manually defined rules
Huang et al. [20]	2013	Robust against large variations in text font, color, size, and geometric distortion	Fails when the text region is partially occluded by other structures
Huang et al. [23]	2014	Computationally fast	Generate a large number of non-text components leading to high ambiguity between text and non-text components

4 Datasets and Protocols

Datasets provides a great help for the successful implementation of the algorithm and achieving the comparison results among various algorithms. Datasets and evaluation methods drive the recent progress in the field of scene text detection and recognition. This section defines the datasets and protocols used to evaluate performance in scene text detection and recognition.

4.1 Datasets

ICDAR 2003: The ICDAR 2003 [24] robust reading competitions were organized in 2003. The total number of fully annotated text images included in this dataset is 509, out of which 258 images are used for training and 251 images are used for testing.

   ICDAR 2011: The ICDAR 2011 [25] robust reading competitions were held in 2011, and it includes the recent advances in the field of scene text detection and recognition. The benchmark used previously in ICDAR competitions is inherited in this datasets but with further extension and modification, as there are many issues with the previous dataset.

4.2 Protocols

Protocols designed for algorithms based on text detection

Precision, Recall, and F_measure are the three important matrices which are used to evaluate the performance of the methods designed for detection of texts. Precision evaluates the ratio between the true positives and all detections. The ratio between the true positives and all true texts that should be detected is evaluated by recall. F_measure is a metric that overall measures the algorithm performance.

Protocol for ICDAR 2003

The parameter **s** is known as match among two rectangles and is given by calculating ratio between intersected areas of rectangles to that of the rectangle which is minimally bounded containing both the rectangles. Rectangles illustrated by every algorithm are known as *estimates,* and ICDAR dataset provides the group of ground true rectangles are known as *targets*. The match with the largest value is calculated for each rectangle. Therefore, for a rectangle q in a set of rectangles Q, the best match s is calculated as:

$$s(q; Q) = \max\{s(q, q' | q' \in Q)\}. \tag{1}$$

Then, we can describe precision and recall as:

$$\text{Precision} = q \in (q; E)/|G|, \tag{2}$$

$$\text{Recall} = \sum q_e \in Es(q_e; G)/|T|, \tag{3}$$

where G and E are the sets of ground true bounding boxes and estimated bounding boxes, respectively. F_measure (F) combines the two defined measures, precision and recall. Parameter β is used to control the relative weights of precision and recall, which is generally set to 0.5 so that the two measures precision and recall gain equal weights:

$$F\_\text{Measure} = \frac{1}{\frac{\beta}{\text{precision}} + \frac{1-\beta}{\text{recall}}} \tag{4}$$

Table 3 illustrates the outcomes of various text recognition schemes calculated over ICDAR 2003 data. One can observe here that the algorithm proposed by Huang et al. [20] performs better than other methods.

Protocol for ICDAR 2011

The computation scheme utilized in ICDAR 2003 and 2005 faces one problem that it is incapable to deal with the cases such as one-to-many and many-to-many matches. In order to deal with this problem, ICDAR 2011 [25] used the evaluation

Table 3 Quantitative comparison of various techniques computed on ICDAR 2003

Proposed schemes	Precision	Recall	F_measure
Huang et al. [20]	**0.81**	**0.74**	**0.72**
Epshtein et al. [17]	0.73	0.60	0.66
Yi et al. [19]	0.71	0.62	0.62
Chen et al. [6]	0.60	0.60	0.58

Table 4 Quantitative comparison of various techniques computed on ICDAR 2011

Proposed schemes	Precision	Recall	F_measure
Huang et al. [23]	**0.88**	**0.710**	**0.780**
Neumann et al. [27]	0.854	0.675	0.754
Huang et al. [20]	0.820	0.750	0.730
Kim et al. [25]	0.830	0.625	0.713
Yi et al. [28]	0.672	0.581	0.623
Yang et al. [25]	0.670	0.577	0.620

scheme developed by Wolf et al. [26]. Definitions of recall and precision are described as below:

$$\text{Precision}(G, N, s_r, s_p) = \sum_j \text{Match}_N(N_j, G, s_r, s_p)/|N| \tag{5}$$

$$\text{Recall}(G, N, s_r, s_p) = \sum_i \text{Match}_G(G_i, N, s_r, s_p)/|G| \tag{6}$$

where N represents the set of detection rectangle and G represents the set of ground true rectangle set. $s_r \in [0, 1]$ denotes the limitation over recall area calculated, and $s_p \in [0, 1]$ denotes area precision limitation. Constraints s_r and s_p are set to 0.8 and 0.4.

Match$_N$ and Match$_G$ are match functions which considers various types of matches. Match$_N$ and Match$_G$ are described as below:

$$\text{Match}_N(N_j, G, s_r, s_p) = \begin{cases} 1 & \text{if one-to-one match} \\ 0 & \text{if no match} \\ U_{sc}(v) & \text{if many } (\rightarrow v) \text{ matches} \end{cases} \tag{7}$$

$$\text{Match}_G(G_i, N, s_r, s_p) = \begin{cases} 1 & \text{if one-to-one match} \\ 0 & \text{if no match} \\ U_{sc}(v) & \text{if many } (\rightarrow v) \text{ matches} \end{cases} \tag{8}$$

where $U_{sc}(v)$ is a parameter function which controls the case of scattering, i.e., splits or merges. The value of parameter $U_{sc}(v)$ is set to 0.8.

Table 4 illustrates the outcomes of various text recognition schemes computed over ICDAR 2011 dataset. We can conclude from table that the performance of the

Fig. 7 a Graph for
F_measure computed on
ICDAR 2003 dataset. **b** Graph
for F_measure computed on
ICDAR 2011 dataset

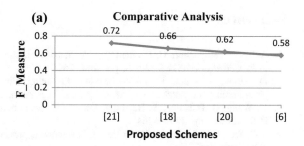

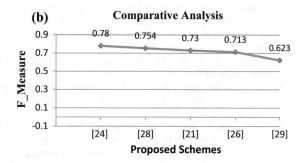

technique given by Huang et al. [23] is the most efficient in comparison with other existing methods. Figure 7 shows the comparative analysis of the existing schemes computed over the ICDAR 2003 and ICDAR 2011 dataset, respectively.

5 Conclusions

This paper presents a through literature review on text recognition in scene images. Component and region-based techniques are mainly focused in this paper. This paper also presents the comparative and quantitative analysis of recent existing schemes based on above-mentioned approaches. In theoretical analysis, various advantages and limitations of techniques are presented. In case of quantitative analysis, various quantitative measures such as recall, precision, and F_measure are used. Through this review work, one can observe that connected component-based methods are more effective to fulfill the requirements of text recognition field. This will also help to the researchers working in the concerned area.

Acknowledgements Authors would like to thank the anonymous reviewers for their valuable suggestions and remarks that helped in improving the quality of the paper.

References

1. A. Mishra, K. Alahari, C. Jawahar, Top–down and bottom–up cues for scene text recognition, in *Proceedings of CVPR* (2012), pp. 2687–2694
2. D. Smith, J. Field, E. Learned-Miller, Enforcing similarity constraints with integer programming for better scene text recognition, in *Proceedings of CVPR* (2011), pp. 73–80
3. K. Wang, B. Babenko, S. Belongie, End-to-end scene text recognition, in *Proceedings of ICCV* (2011), pp. 1457–1464
4. Y.C. Wei, C.H. Linm, A robust video text detection approach using SVM. Expert Syst. Appl. 10832–10840 (2012)
5. J. Weinman, E. Learned-Miller, A. Hanson, Scene text recognition using similarity and a lexicon with sparse belief propagation. IEEE Trans. PAMI 1733–1746 (2009)
6. X. Chen and A. Yuille, Detecting and reading text in natural scenes, in *Proceedings of CVPR* (2004)
7. Y. Zhong, K. Karu, A.K. Jain, Locating text in complex color images. Pattern Recogn. **28** (10), 1523–1535 (1995)
8. K.I. Kim, K. Jung, J.H. Kim, Texture-based approach for text detection in images using support vector machines and continuously adaptive mean shift algorithm. IEEE Trans. PAMI **25**(12), 1631–1639 (2003)
9. J. Gllavata, R. Ewerth, B. Freisleben, Text detection in images based on unsupervised classification of high-frequency wavelet coefficients, *in Proceedings of ICPR* (2004)
10. H.P. Li, D. Doermann, O. Kia, Automatic text detection and tracking in digital video. IEEE Trans. Image Process. **9**(1), 147–156 (2000)
11. M.R. Lyu, J. Song, M. Cai, A comprehensive method for multilingual video text detection, localization, and extraction. IEEE Trans. CSVT **15**(2), 243–255 (2005)
12. P. Viola, M. Jones, Fast and robust classification using asymmetric adaboost and a detector cascade, in *Proceedings of NIPS* (2001)
13. S.M. Lucas, Text locating competition results, in *Proceedings of ICDAR* (2005)
14. V. Wu, R. Manmatha, E.M. Riseman, Finding text in images, in *Proceedings of 2nd ACM International Conference Digital Libraries* (1997)
15. C. Wolf, J.M. Jolion, Extraction and recognition of artificial text in multimedia documents. Formal Pattern Anal. Appl. **6**(4), 309–326 (2004)
16. K.Wang, S. Belongie, Word spotting in the wild, in *Proceedings of ECCV* (2010)
17. B. Epshtein, E. Ofek, Y. Wexler, Detecting text in natural scenes with stroke width transform, in *Proceedings of CVPR* (2010)
18. C. Yao, X. Bai, W. Liu, Y. Ma, and Z. Tu, Detecting texts of arbitrary orientations in natural images, in *Proceedings of CVPR* (2012)
19. C. Yi, Y. Tian, Text string detection from natural scenes by structure-based partition and grouping. IEEE Trans. Image Process. **20**(9), 2594–2605 (2011)
20. W. Huang, Z. Lin, J. Yang, J. Wang, Text localization in natural images using stroke feature transform and text covariance descriptors, in *Proceedings of ICCV* (2013)
21. L. Neumann, J. Matas, A method for text localization and recognition in real-world images, in *Proceedings of ACCV* (2010)
22. P. Shivakumara, T.Q. Phan, C.L. Tan, A laplacian approach to multi-oriented text detection in video. IEEE Trans. PAMI **33**(2), 412–419 (2011)
23. W. Huang, Y. Qiao, X. Tang, Robust scene text detection with convolution neural network induced mser trees, in *Proceedings of ECCV* (2014)
24. S.M. Lucas, A. Panaretos, L. Sosa, A. Tang, S. Wong, R. Young. ICDAR 2003 robust reading competitions, in *Proceedings of ICDAR* (2003)
25. A. Shahab, F. Shafait, A. Dengel. ICDAR 2011 robust reading competition challenge 2: reading text in scene images, in *Proceedings of ICDAR* (2011)

26. C. Wolf, J.M. Jolion, Object count/area graphs for the evaluation of object detection and segmentation algorithms. IJDAR **8**(4), 280–296 (2006)
27. L. Neumann, J. Matas, On combining multiple segmentations in scene text recognition, in *Proceedings of ICDAR* (2013)
28. C. Yi, Y. Tian, Text detection in natural scene images by stroke gabor words, in *Proceedings of ICDAR* (2011)

Novel Approach of Spoofing Attack in VANET Location Verification for Non-Line-of-Sight (NLOS)

Babangida Zubairu

Abstract Location verification of a vehicle in VANET under non-line-of-sight (NLOS) condition requires the questioned vehicle to be cooperatively located by questioner via another vehicle that has direct link with the vehicle in a state of NLOS condition. One of the challenges is to prove that the vehicle is physically located in a claimed position, and MHLVP has addressed this issue with short-coming in security like spoofing attack. In this paper, vital risks associated with vehicle location verification in VANET that could hinder the performance and efficiency of MHLVP were identified and illustrated, and also solutions to the risks were presented; hence, developing secured and reliable protocol is the backbone of modern intelligence transportation system (ITS).

Keywords VANET · NLOS · ITS · MHLVP

1 Introduction

The rate at which competition exists amongst technological industries and necessity to increase road safety on roads have led to several innovations and paradigm shift in modern transportation system of this present generation and beyond [1], vehicles are becoming "computers on wheels", or rather "computer networks on wheels" [2]. Various research efforts from automotive industry and academia are engaged to speed up the deployment of a various wireless technology for communications and information exchange amongst vehicles moving on the road and roadside infrastructure [3]. Vehicular ad hoc network (VANET) offers safety to driver and passengers; it also provides other applications that are useful as well; as such VANET is considered to be a vibrant factor of the modern future of ITS. VANET is symbolized by some features such as higher speed, limited connectivity and rapidly

---

B. Zubairu (✉)
School of Computer & System Science, Jaipur National University,
Jaipur, Rajasthan, India
e-mail: zaibag999@gmail.com

© Springer Nature Singapore Pte Ltd. 2018
B. Panda et al. (eds.), *Innovations in Computational Intelligence*, Studies in
Computational Intelligence 713, https://doi.org/10.1007/978-981-10-4555-4_4

varying topologies of the vehicles [4]. The vehicles present mobility nodes of hundreds of kilometres per hour, as such the number of vehicles changes intensely in a bulky scale which requires frequent invoking of position authentication of vehicles and constant updating of information such as vehicle position information, thus, gaining a large overhead of communication and control [5]. VANET is emerging as the most challenging technology integrating ad hoc networks [6]. It allows vehicles to communicate directly with each other known as vehicle-to-vehicle communication (V2V). Vehicles can communicate with roadside units via vehicle-to-infrastructure communication (V2I) [7], and the roadside unit (RSU) infrastructure connects the vehicles to a control system and the Internet [8] via trusted authority (TA) [9]; application servers were also installed at the back end. The vehicles are prepared with an on-board units (OBUs) for vehicles to communicate amongst themselves and with the RSUs. The OBU consists of processors as well as sensors for information exchange, global positioning system (GPS) which helps in vehicles direction and position. Event data recorder (EDR) is also attached to the vehicles [8]. The communications amongst vehicles are exposed to signal interference since they move in a varied environmental setting and conditions; however, objects on the roadside such as buildings, trees, moving trucks and construction on the sides of the road and location landscape can hinder the radio signals as well as halt proper communication amongst vehicles, and these could lead to vehicle communication and signal line of sight be obstructed by generating NLOS. As result of this, drivers could make erroneous decision while moving on the road and change wrong path [10]. Attack like **spoofing on one of the communicating vehicles in which one vehicle could successfully masquerades as another by falsifying and injecting wrong information thereby gaining an illegitimate advantage** can serve as an impediment amongst vehicles within the same communication range when there is NLOS. Hence, safety and information exchange protection in VANET are essential since it affects the life of people, and the messages sent between vehicles need to be authenticated and carried to the authorized vehicles as quickly as possible [11]. VANET becomes target of most of the attacks such as spoofing due to its openness nature, where adversaries alter node identity and communicating information amongst the vehicles [12]. Research in VANET as expressed in [13] attracts the attention of researchers from automotive industries to academia across the globe working tirelessly towards making VANET a reality as a modern intelligence transportation system (ITS).

2 Review of Literature

Various works have been done and still ongoing by researchers to develop a system that can protect position of devices' information in wireless and ad hoc networks. Hence, the networks have a different features and requirements, and a number of security structures and solutions were proposed for different networks settings, Watfa in [14] view VANET as a network characterized by the overlapping nature of information transmission as a result of higher mobility of vehicles; the rate of speed at which vehicles move; the dynamic nature of vehicles movement and the operating environment. VANET challenges differ from mobile ad hoc network (MANET) due to higher mobility rate of vehicles moving on a different directions and communicating within the same range, and these highlighted that the higher mobility of vehicles in VANET the more routing problem [15]. However, related problems have been addressed in recent work of [10] and [16] the authors proposed location verification amongst cooperative neighbouring vehicles to overcomes NLOS effects. The work highlighted collaborative approach to verify an announced location when straight communication between the questioned vehicle and the verifier is not feasible, and the work presented by [16] follows the same approach of [10], though some drawbacks were observed with the work. Firstly, the simulation results highlighted the protocol's softness to attacks like message forgery and false requests. Secondly, the solution assumes that there is at least one joint neighbour vehicle that has straight connection with the requester and questioned vehicle, as observed multi-hop location verification protocol helps a vehicle to verify its neighbours when direct communication path between verifier and claimer vehicles is obstructed as a result of NLOS; thus the vehicle location information can be attacked by injecting false information into a vehicle from an outsider or malicious attack from insider on the communicating vehicles [17], and this type of attack could affect the privacy and integrity of vehicle's location information.

In [18], support vector machines (SVMs) was proposed as a method for improving the accuracy and determining the number of attackers using spatial information for detecting spoofing attacks as well as the number of attackers and localizing them. In [19] aware service discovery mechanism based on geographical location for vehicle was proposed, the mechanism takes advantage of IPv6 multicast service discovery protocol GeoNe, though, improvements are needed in the multi-hop scenario, also additional mechanisms are necessary such as message confidence values which can be gained using cryptographic mechanisms, approach presented in [18, 20]. In [21], new approach was proposed in location verification of neighbour in ad hoc network by integrating the neighbour position verification protocol with spontaneous ad hoc network protocol in which node in the network verifies the position of its neighbours as well as performing security analysis of the system.

The work presented in [10] and [16] was based on location confirmation of vehicles and an approach to overcome the effects of NLOS in VANET. The protocols use cooperative approach to authenticate a questioned vehicle and its proclaimed position when straight link is not feasible, and the simulation results show softness of the protocols on attacks like spoofing (Fig. 1).

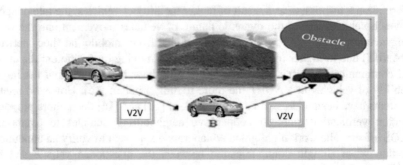

Fig. 1 Picture depicting obstacle in vehicle location

Fig. 2 Vehicle in NLOS
state graph

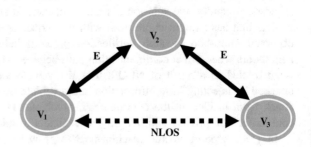

3 Motivation and Problem Statement

In designing VANET, fixing of obstacles effects should be considered; the obstacles
could be buildings, topography of the area and as well as the objects that moves on
the road, and all these could lead to signal blockage. Moreover, addressing challenges associated with attack like spoofing in NLOS condition is also a paramount
importance.

Let $G = (V; E)$ be a graph, V represents the set of vehicles(vertices) and
E represents the set of links(edges) between a vehicle to another vehicle, the
vertices V_1 to V_3 of graph G are disconnected as shown in Fig. 2.

By definition, not all vertices of G are joined by paths to V_i. Vertex V_i and all
the vertices of G that have path to V_i together with the entire edges incidence on
them form the components of the graph.

3.1 Lemma

The graph G is said to be disconnected if and only if all its vertices set of V can be
partitioned into two non-empty separate subsets V_1 to V_2, V_2 to V_3 ∃ there exist no
edge(link) in G whose one end of the vertices is in subset V_1 to V_3.

3.2 Proof

Suppose that such a splitting exists, consider two arbitrary vertices, V of G, $\exists$ $a \in V_1$ and $b \in V_2$; no path exists between vertices V_1 to V_3, otherwise, there would be at least one path whose one end vertex would be in V_1 to V_2 and V_2 to V_3. Hence, if a barrier occurs, then G is not connected.

Similarly, let G be a disjointed graph; consider a vertex V_i in G. Let V_i be the set of all vertices that are joined by link to another. Since G is disjointed, V_1 is joined to any in V_2 and V_2 to V_3 by an edges but no path exist between V_1 to V_3, hence the partitioned.

4 Methodology

4.1 Multi-hop Location Verification Protocol (MHLVP) Overview

The protocol developed in [10] was able to prove proclaimed position of a questioned vehicle using a cooperative multi-hop approach, when direct connection is not conceivable between questioned vehicle and verifier; the protocol was based on assumptions as follows:

- The mobility and location of the vehicles are known by all the vehicles that are within the same communication range.
- The presence of line of sight amongst atleast two vehicle, each vehicle can verify its neighbour vehicle with the help of receive signal strength and calculate its neighbour's distance.
- The vehicles can communicate amongst themselves and share resources with the help of line of sight.

4.2 Position Verification Computation of MHLVP

Triangular coordinates system approach was employed to calculate the position of a vehicle in questioned, as illustrated in Fig. 3, vehicle A wants to validate vehicle C's position; however, straight connection is not feasible due to the presence of abstraction from the topography of the area, though vehicle B can link straight with both vehicles A and C (Fig. 3).

The formula proposed for obtaining distance between verifier and questioned vehicle is given as:

Fig. 3 Obstruction in vehicles communication

$$d_{ac} = \sqrt{d_{bc}^2 + d_{ab}^2 - 2d_{bc}d_{ab}\cos\theta} \qquad (1)$$

where $\theta = \arccos(\vec{BA}.\vec{BC})$,

$\vec{BA}$ is the distance between vehicle B to A and

$\vec{BC}$ is the distance between vehicle B to C

5 Potential Issues with MHLVP

The results of the protocol's simulations show significant improvement of vehicle location verification in channel capacity utilization, message delivery success rate but softness of the protocol on security attacks like spoofing, these issues have raised some questions that need to be addressed as illustrated in scenario 1 and 2.

5.1 Scenario 1

For the this scenario, existence of malicious vehicles amongst the communicating vehicles is considered, and the malicious vehicles performed different attacks that might affect the protocol's performance with some possible attacks such as message fabrication and wrong requests, see Fig. 4.

In the above scenario, vehicle A wants to prove vehicle **C**'s position; but direct communication is not feasible due to the presence of hitches (NLOS). Therefore, vehicle **A** initiates the location verification discovery process, and vehicle **D** claims to have an active link to **C** and pretends that it must be next vehicle to be contacted not any other vehicle if vehicle **A** wants to send location verification request of vehicle **C**.

This kind of attack could be dangerous because a vehicle can claim to be in diverse points at the same time, thereby generating anarchy and vast security network threats. It could also lead to the damages of network topology and links as well as unexpected consumption of network bandwidth, and this implies that we not only need protection of message exchange contexts but also we need secure system that can enable the detection of attack from unauthorized vehicles and alleviate them from the network.

5.2 Scenario 2

An insider attack is considered from a malicious vehicle; the vehicle successfully masquerades as another by falsifying data and thereby gaining an illegitimate advantage. To illustrate this threat, consider vehicle **A** wishes to prove vehicle **C**'s position; however, direct communication is not feasible due to the presence of

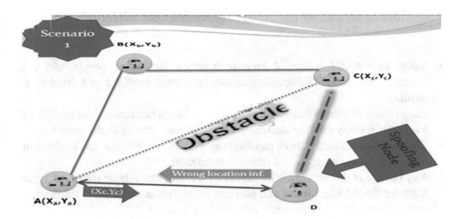

Fig. 4 Spoofing attack in MHLVP of scenario 1

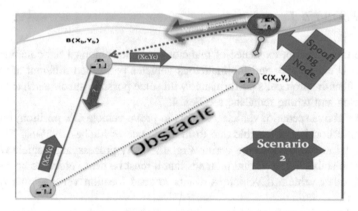

Fig. 5 Spoofing attack in MHLVP of scenario 2

hitches, but vehicle **B** can connect directly with vehicles **A** and **C**; however, when vehicle **B** is sending message to verify vehicle **C**'s location, the message is intercepted by spoofing vehicle S_i, and the vehicle **B** is injected with false information about vehicle **C**, with the aim to disrupt the communication and messages exchange amongst the vehicles (Fig. 5).

In this scenario, vehicle B is fabricated with wrong positioning of vehicle **C** from a malicious vehicle. Hence, the *protocol should be able to identify whether the information about a vehicle is compromised or not*. Communications passing through a vehicular network as well as information about the vehicles location must be secured and protected to ensure the smooth functioning of intelligent transportation systems.

This attack can course tragic costs such as the losses of lives or loss of time such as altering traffic jam as a result of the spoofing attack.

5.3   Consequences of the Spoofing Attacks on MHLVP

- Safety in VANET is crucial because it affects the life of people, and it is essential that the vital information cannot be altered, modified or deleted by an attacker.
- These kinds of attacks are dangerous since a vehicle can claim to be in different locations, thereby causing anarchy and big security threats in the network.
- The above-mentioned threats can damage VANET connection and as the same time increase consumption of network bandwidth.
- Any change in the network information by a fraudulent vehicle may cause great harm for the vehicle, drivers and passengers, as well as partitioning the network and decreasing the performance of the whole network.

6   Secured Position Verification

To address the mentioned threats, I proposed a working model with three-level approaches where vehicles that have direct link with vehicle in NLOS collaborate to determine presence of the spoofing attacker, its location and facilitate localization of the attacker in sequential manner based on the following mechanisms:

- Check for existence of spoofing vehicle within a communication range.
- Determine the location of the spoofing vehicle if any.
- Localize spoofing vehicle by getting rid of them from the network (Fig. 6).

The model above depicts the process of spoofing attack detection, location and localization in MHLVP; hence it is important to secure location verification of vehicle in NLOS condition against spoofing attack.

6.1   Detection Process

The detection process is based on temporally ordered routing algorithm (TORA), and the protocol finds and maintains multiple routes to the destination and reacts

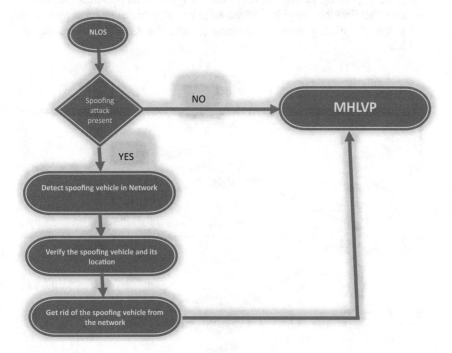

Fig. 6 Model depicting the process of spoofing attack detection, location and localization in MHLVP

only when all routes to the destination are lost; to establish a route, the vehicle first broadcasts a request packet to its neighbours. This request is rebroadcasted throughout the network until it reaches an intermediate vehicle that has path to the destination of the vehicle in questioned. The recipient of the request packet then broadcasts the update packet which lists its distance with respect to the destination of the vehicle [22], and the localization approach provides real-time attack protection of vehicle position in NLOS condition. The overall detection processes performed by each vehicle were classified into three stages as follows (Fig. 7):

- **Stage 1**: $Node_i$ sends broadcasts beacon messages to its neighbours to verify the location of a vehicle in NLOS state and waits to receive beacon messages reply from neighbouring vehicles. At this stage, $node_i$ is an initiator, and it serves as claimer and verifier. When the vehicle sends a request message to its neighbours, it is serving as a claimer function, and it serves as verifier function when it receives reply from its neighbours. Each vehicle can play all these roles at various moments and for different purposes.
- **Stage 2**: When $node_i$ receives locations and strength measurements of the vehicle in NLOS through its neighbours, $node_i$ performs verification of the results obtained; and the distance between initiator to verifier and verifier to NLOS vehicle. If inconsistency detects the enhanced position, verification of algorithm on the vehicles would be performed.
- **Stage 3**: If spoofing vehicle proves to be in stage 2, $node_i$ performs the spoofing vehicle classification algorithm on the neighbours' vehicles and attempts to find all potential spoofing vehicles originating from the same malicious physical vehicle.

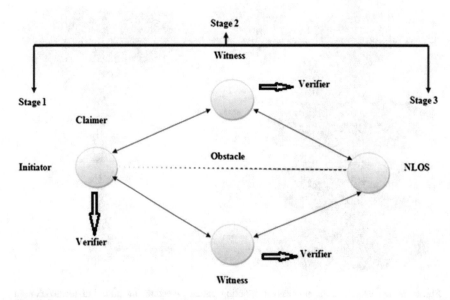

Fig. 7 Detection process

- **During Stage 3**, the estimated physical position of the malicious vehicle would be identified and even its movement route, which would be helpful for further intrusion response decisions and detection.

6.2 Vehicle Functions

Initiator: This vehicle can serve as a claimer and verifier as follows:

- Claimer function. The $node_i$ sends request message to its neighbours (witness) to determine location of a $node_j$ in NLOS condition. The request message contains its announced position (x_c, y_c) and mobility vector (Fig. 8).
- Verifier function: Verification function is performed at two levels; the witness level and initiator level with aim of maintaining consistency of messages amongst communicating vehicles that are within the same communication range. The verification model is given below (Fig. 9).

Verification at witness level: In the above model, two levels of verifications were proposed. The first is at witness vehicles; this vehicle will use locations techniques to determine the location of attacker. The techniques proposed are the time difference, as signal travels in constant speed and the difference in arrival times will be used to predict distance of an attacker with the help of timestamps in the packet. The second is round-trip time (RTT); this method uses the time required for the packet to travel from a precise source to a precise destination, and back again, so the change in propagation time will be used to determine the distance of the attacker. The third is received signal strength indication (RSSI); this will be used to measure the distance between two or more vehicles using network interface card (NIC), since network card provides interface to RSSI.

Verification at initiator level: At this level, attack detector was defined and formulated using statistical test with the aim of identifying the presence of a spoofing vehicle in VANET under NLOS condition.

Statistical test: In the statistical test, the null hypothesis is: H_0; normal and alternate hypothesis is H_1; attack presence in the network. To evaluate whether calculated data from witness vehicles belong to the null hypothesis or else, test statistic Z is used, and when the calculated test statistic Z^{cal} varies from the hypothesized values, the null hypothesis is rejected and claimed the existence of vehicle performing spoofing attack [23]; hence, spoofing attack detected (Fig. 10).

Initia-   *Witness*

Fig. 8 Vehicle functions

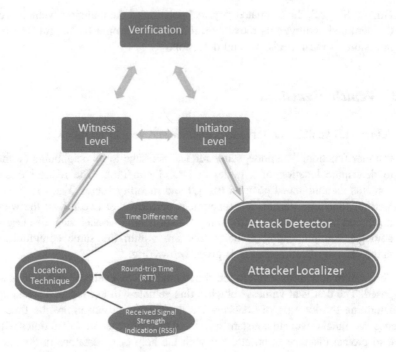

Fig. 9 Verification model

Fig. 10 Detection model

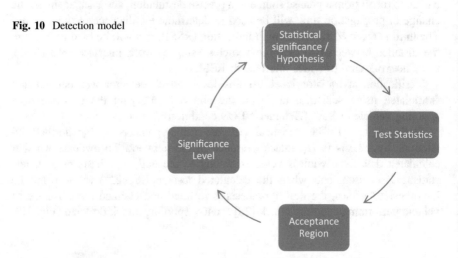

- Statistical significance testing
  Null hypothesis:
  H_0: Attacks not present (normal)
  Alternate hypothesis:
  H_1: Attack present (not normal)

- Test statistic Z
  Z = Normal Distribution for received replies from witness vehicles.

$$Z = \frac{\bar{X} - \mu}{\sigma/\sqrt{n}} \tag{2}$$

$\bar{X}$ Mean of sample replies from witness vehicles
μ Mean of the received replies from witness vehicles
σ Standard deviation of received replies from witness vehicles
n Total number of witness vehicles

Acceptance region $\mathbf{\Omega}$

If $\mathbf{Z^{cal}} \in \mathbf{\Omega}$, no attack (normal)
If $\mathbf{Z^{cal}} \notin \mathbf{\Omega}$, attack is present

Significance testing with significance level α
The probability of rejecting the null hypothesis given that it is true of the statistical test at 5% level of significance
Localizer: The training data for the Z statistics are the random reply received from the witness vehicles within a specified time period. The training data are to be tested whether an attack occurs or not; if an attack occurred, then triangular inequality is embarked to identify the vehicle performing the attack amongst the communicating vehicles. Triangular inequality is proposed to get rid of the malicious vehicle out of the network.
Theorem: If A, B, C and D are arbitrary nodes with triangle ADC, the distance from node A to D as $\mathbf{d_{ad}}$ and distance of node D to C as $\mathbf{d_{dc}}$, then

$$|d_{ad} + d_{dc}| \le |d_{ad}| + |d_{dc}| \tag{3}$$

Hence, triangle inequality states that for any given triangle ACD, the sum of the lengths of any two sides of the triangle (i.e. AD + DC) must be greater than or equal to the length of the other remaining side (AC).
In normal setting, the sum of the distance between the initiator to witness and witness to NLOS vehicle should be more than or equal to the distance between initiator to NLOS vehicle. If the computed distance between initiator and NLOS vehicle is greater than the sum of the distance between initiator to witness and witness to NLOS vehicle, then the witness vehicle is not an active vehicle, and it will be get rid out of the network.

7 Conclusion

Security and privacy are the major challenges in VANET position verification for deployment of modern intelligence transportation system. In this work, spoofing attack has been identified as major drawback from the work done by [10, 16], and they presented a collaborative approach in vehicle location verification in multi-hop approaches. In MHLVP, vehicles prove neighbours position when straight connection between verifier and claimer is obstructed due to NLOS; the system shows softness of the protocol to security attacks. Spoofing attack has been identified as a major threat in MHLVP approach, and solution to this threat has been proposed that can locate, detect and localize malicious vehicle amongst communicating vehicles from VANET network.

References

1. P. Papadimitratos, A. De La Fortelle, K. Evenssen, R. Brignolo, S. Cosenza, Vehicular communication systems: enabling technologies, applications, and future outlook on intelligent transportation. IEEE Commun. Mag. **47**(11), 84–95 (2009)
2. M.H. Kabir, Research issues on vehicular ad hoc network. Int. J. Eng. Trend Technol. **6**(4), 177–179 (2013)
3. A. Ghosh, V.V. Paranthaman, G. Mapp, O. Gemikonakli, Exploring efficient seamless handover. EURASIP J. Wirel. Commun. Netw. **1**(227), 1–19 (2014) [Online]. http://jwcn.eurasipjournals.springeropen.com/articles/10.1186/1687-1499
4. R.S. Raw, M. Kumar, N. Singh, Security challenges, issues and their solutions for VANET. Int. J. Netw. Secur. Appl. (IJNSA) **5**(5), 95–105 (2013)
5. G. Xue, Y. Luo, J. Yu, M. Li, A novel vehicular location prediction based on mobility patterns for routing in urban VANET, EURASIP J. Wirel. Commun. Netw. **222**(1), 2012–2222 (2012) [Online]. http://jwcn.eurasipjournals.springeropen.com/articles/10.1186/1687-1499
6. A.K. Dhami, N. Agarwal, Challenges in securing VANET: the intelligent transportation system. Int. J. Comput. Sci. Secur. (IJCSS) **6**(6), 366–375 (2012)
7. H. Vahdat-Nejad, R. Azam, M. Tahereh, M.W. Mansoor, A survey on context-aware vehicular network applications. Elsevier: Veh. Commun. **3**(1), 43–57 (2016)
8. L. Wischhof, A. Ebner, H. Rohling, Information dissemination in self-organizing intervehicle networks, Intelligent transportation systems. IEEE Trans. **6**(1), 90–101 (2005)
9. N. Jayalakshmi, Designing of regional trusted authority with location based service discovery protocol in VANET. Int. J. Res. Appl. Sci. Eng. Technol. (IJRASET) **1**(1), 1–8 (2013)
10. O. Abumansoor, B. Azzedine, A secure cooperative approach for nonline-of-sight location verification in VANET. IEEE Trans. Veh. Technol. **61**(1), 275–285 (2012)
11. K. Lim, D. Manivannan, An efficient protocol for authenticated and secure message delivery in vehicular ad hoc networks. Elsevier Inc: Veh. Commun. **4**(1), 30–37 (2016)
12. R.M. Shingade, V.A. Abhishek, Detection of spoofing attackers in wireless network. Int. J. Innovative Res. Adv. Eng. **1**(4) (2014)
13. H. Hartentein, K. Laberteaux, VANET: *Vehicular Application and Inter-Networking Technology*, 2nd edn. (Wiley, Ltd, publishers, United Kingdom, 2010)
14. M. Watfa, *Advances in vehicular ad-hoc networks: developments and challenges*, 1st edn. (Information Science Reference IGI Global, New York, USA, 2010)

15. C. Falko, D Summer, *Vehicular Networking*, 1st edn. (Cambridge University Press, Cambridge, 2015)
16. G.J. Archanaa, R. Venittara, A MultiHop location verification protocol for NLOS condition in VANETs. Int. J. Eng. Sci. Comput. (IJESC) **10**(4), 476–480 (2014)
17. A.M. Malla, R.K. Sahu, Security attacks with an effective solution for DOS attacks in VANET, Int. J. Comput. Appl. **66**(22), 45–49 (2013)
18. M. Barapatre, V. Chole, L. Patil, A review on spoofing attack detection in wireless adhoc network. Int. J. Emerg. Trends Technol. Compu. Sci. (IJETTCS) **2**(6), 192–195 (2013)
19. S. Noguchi, T. Manabu, E. Thierry, I. Astuo, F. Kazutoshi, Design and field evaluation of geographical location-aware service discovery on IPv6 Geo networking for VANET. EURASIP J. Wirel. Commun. Netw. **29**(1), 1–16 (2012)
20. J.N. Al-Karaki, E.A. Kamal, Routing techniques in wireless sensor networks: a survey. IEEE Wirel. Commun. **11**(6), 6–28 (2004)
21. K. Padmavathi, L. Jaganraj, A study on secure spontaneous ad hoc network protocol for neighbor position verification. Int. J. Adv. Res. Comput. Sci. Softw. Eng. (IJARCSSE) **3**(10), 1187–1192 (2013)
22. H.P. Ambulgekar, S.H. Raut, Proactive and reactive routing protocols in multihop. Int. J. Adv. Res. Comput. Sci. Softw. Eng. **3**(4), 152–157 (2013)
23. W. Trappe, R.P. Martin, Chen Y, in *Detecting and Localizing Wireless Spoofing Attacks*. 4th Annual IEEE Communications Society Conference on Sensor, Mesh and Ad Hoc Communications and Networks, San Diego, CA, 2007, pp. 193–202

An Enhanced Method
for Privacy-Preserving Data Publishing

Shivam Agarwal and Shelly Sachdeva

Abstract A wide collection of sensitive information is available on the Internet. Privacy of this released data needs to be maintained. It is one of the biggest challenges in an information system. It is very essential to preserve the publishing data. The privacy in publishing data is applied before releasing the data. The techniques used for preserving privacy of published data are k-anonymity, l-diversity, and t-closeness. But these techniques have some limitations. K-anonymity prevents from the record linkage attack but fails to prevent attribute linkage attack. L-diversity overcomes the limitation of k-anonymity technique. But, it cannot prevent address identity disclosure attack and attribute disclosure attack in some exceptional cases. T-closeness technique preserves the attribute linkage attack but does not identity disclosure attack, but it has large computational complexity. In all the techniques, there is one more issue named as data utility. Thus, in current paper, we propose a new technique which is an enhanced version of k-anonymity. An algorithm is presented named as (P, U)-sensitive k-anonymity, which minimizes the rate of attacks and increases the data utility of published data. This technique gives us better result in terms of running time, similarity attack, and data utility.

Keywords Sensitive attributes · Datasets

1 Introduction

Published data is the microdata which is published by the government and private sector for the purpose of research. This data may contain the information about individuals and other things. It may contain some sensitive information such as

S. Agarwal (✉)
Department of Computer Science, Pranveer Singh Institute of Technology, Kanpur, India
e-mail: agarwal.shivam88@gmail.com

S. Sachdeva
Department of Computer Science, Jaypee Institute of Information Technology, Noida, India
e-mail: sachdevashelly1@gmail.com

© Springer Nature Singapore Pte Ltd. 2018
B. Panda et al. (eds.), *Innovations in Computational Intelligence*, Studies in
Computational Intelligence 713, https://doi.org/10.1007/978-981-10-4555-4_5

diseases and salary, and some personal information. No individual wants to disclose his/her sensitive information. The key problem of data is confidentiality of sensitive information in publishing data. There is a need to preserve the data before publishing to the people. The leakage of data can lead to remarkable harm (damage) to individual both emotionally and/or materially. The data publishing need both the data integration and data sharing in such a manner so that there is no privacy loss, but it is very typical to develop some techniques which enable both the integration and sharing of data without losing privacy [1]. The published data is in tabular form, which contains two types of attributes, one is public attribute and another is private attribute. The public attribute is also known as quasi-identifier (QI). With the help of quasi-identifier, we can retrieve the information of a particular individual. These are publicly known. The private attributes are the sensitive attributes, which contains the sensitive information of individuals. So there is a need to secure them. To provide security to sensitive attributes, we use suppression [2] and generalization [3]. These are applied only on the quasi-identifiers. For example, in suppression, we suppressed the numeric data by (*). The value of (*) lies between (0 and 9). Another example to generalize the 'age' attribute whose value is 20 is to write as 'age ≥ 30.' There are so many other ways to generalize the age attribute. Thus, no one is able to know the exact values of the quasi-identifiers, which leads to the leakage of sensitive attributes. For retrieving the sensitive information, linkage attack [2] is used. In linkage attack, there are so many published datasets which are published from a single database, so by linking two or more published dataset a person may recognize the sensitive information of published data. There are two types of linkage attack, one is attribute linkage attack and another one is record linkage attack. A recent survey estimated that 87% of US population is uniquely identified with the help of 'linkage attack' by using the quasi-identifiers such as name, age, and zip code of 5 digits [4]. To prevent the published data from this linkage attack, privacy-preserving techniques are used. The most widely used techniques are k-anonymity [2], l-diversity [3], and t-closeness [5]. But these techniques have some limitations. Some enhanced techniques such as 'P-sensitive k-anonymity' and '(p, α)-sensitive K-anonymity' are used which are used to preserve the data. But they also have some limitations. There is another concept come in this field, i.e., negative data publication [6]; in this to improve the level of privacy, the original sensitive values are replaced by negative values. So this will help us to prevent the identity disclosure attack. In this, the data utility is also maintained.

Data utility [7] is the rate of utilization of data in the released dataset. The utilization of data is defined as the use of data. If the data is much understandable or usable, then the rate of data utility loss is very low. Section 2 contains brief description of techniques which are used to preserve the published data. In Sect. 3, a new technique which is an enhanced version of k-anonymity is proposed. Section 4 details the performance of the algorithm. Finally, in Sect. 5, the conclusion and future directions are given.

2 Privacy-Preserving Data Publishing

There are three existing techniques that are widely adopted for preserving the published data. These methods are k-anonymity, l-diversity, and t-closeness.

Sweeny introduced K-anonymity [2], a property by which a record is indistinguishable from at least $k - 1$ records. In this, if the sensitive value has same values in equivalence class, then privacy cannot be achieved. This technique prevents the dataset from linkage attack. The background knowledge attack and the homogeneity attack are not prevented by this method. Thus, k-anonymity prevents the record linkage attack but cannot prevent the attribute linkage attacks. Basically, there are two main models for k-anonymity: one is global recording [8–10] and another is local recording [10, 11]. In this paper, we use local recording k-anonymity.

L-diversity method [3] overcomes the limitations of K-anonymity. In this method, the sensitive attribute contains l 'well-represented' values. The attribute linkage attack is prevented. However, it cannot prevent the skewness and similarity attack.

In t-closeness method [5], the distance between distribution of a sensitive attribute in a class and distribution of the attribute in the whole table is not more than the threshold. It prevents the data from homogeneity to background knowledge attacks. However, it has large computational complexity. In t-closeness, there exists a trade-off between data utility and data privacy.

The k-anonymity provides some privacy to the released dataset. Still it does not prevent from various attacks. L-diversity provides better security than k-anonymity. However, in t-closeness, the dataset is secure, but the rate of data utility loss [7] is very high. Data utility is the utilization of the data in the released dataset. Suppose in a dataset all the quasi-identifiers are suppressed to (*), then we cannot extract the values to identify the data which means the use of data is negligible. Thus, we can say the rate of data utility loss is very high. So to maintain privacy and utility of the data, we have to focus on two existing techniques, i.e., k-anonymity and l-diversity. Thus, in this paper to minimize the data utility loss, the new technique is proposed which is based on k-anonymity. Table 1 shows 4-anonymous data.

K-anonymity states that in equivalence class, there will be at least $K - 1$ records. Homogeneity attack is performed if the values in the sensitive attribute column are same. This is a privacy breach. Tuples having S.No. 4, 5, 6, and 7 of Table 1 contains same sensitive attribute. Suppose Alice knows that Bob lives in zip code

Table 1 4-anonymous data depicting homogeneity attack

S.No.	Zip code	Age	Nationality	Disease
4	130**	4*	*	Cancer
5	130**	4*	*	Cancer
6	130**	4*	*	Cancer
7	130**	4*	*	Cancer

Table 2 Background knowledge attack on 4-anonymous data

S.No.	Zip code	Age	Nationality	Disease
1	123**	>30	*	Heart disease
2	123**	>30	*	Heart disease
3	123**	>30	*	Viral infection
4	123**	>30	*	Viral infection
5	4567*	≤40	*	Cancer
6	4567*	≤40	*	Heart disease
7	4567*	≤40	*	Heart disease
8	4567*	≤40	*	Flu

Table 3 2-sensitive 4-anonymous microdata

S.No.	Zip code	Age	Nationality	Disease
1	123**	>30	American	Cancer
2	123**	>30	American	Cancer
3	123**	>30	American	HIV
4	123**	>30	American	HIV
5	456**	4*	Indian	Cancer
6	456**	4*	Indian	Flu
7	456**	4*	Indian	Flu
8	456**	4*	Indian	Indigestion

13056 and his age is 42, and now, Alice can easily predict that Bob is a cancer patient. Thus, we can see that homogeneity attack is performed on Table 1.

The background knowledge attack is that the attacker knows some knowledge about the victim, and on the basis of that knowledge, he wants to know the victim information. Table 2 shows how the background knowledge attack is performed.

Suppose Alice knows that Bobby is a female and her nationality is Japanese. He also knows that her age is 21 years. The tuples 1, 2, 3, and 4 satisfy this information. Alice also knows that Japanese has a very low probability of heart disease, so he can easily predict that Bobby has a viral infection. Thus, through background knowledge, he can predict the information of the victim. In order to overcome the drawbacks, there exist algorithms to enhance the k-anonymity such as 'P-sensitive k-anonymity' and $(p - \alpha)$-sensitive k-anonymity.

2.1 P-Sensitive K-Anonymity

The released dataset T satisfies p-sensitive k-anonymity [12] property, if the dataset satisfies the k-anonymity for all the QI group, and in each QI group, the number of distinct values of sensitive attributes is at least p. This technique preserves the dataset beyond k-anonymity and also protects the dataset from attribute disclosure. Consider Table 3 satisfying 4 anonymity. There are two QI groups, one containing tuples 1, 2, 3, and 4, and another containing 5, 6, 7, and 8. In each QI group, the disease column has at least two distinct values. Thus, we can prevent the homogeneity attack. Also, we can prevent from attribute disclosure. But, it has some shortcomings. Suppose QI group one (tuple 1, 2, 3, and 4) has sensitive attribute containing values HIV and Cancer only. These are the most sensitive diseases. The attacker can know that the victim is suffering from very serious disease. This is the drawback of this technique.

2.2 (p, α)-Sensitive k-Anonymity

The released dataset satisfies the (p, α)-sensitive k-anonymity [13], if the dataset satisfies the k-anonymity and each QI group has at least 'p' distinct values of sensitive attributes with the total weight at least equal to α. Let us consider Table 4.

Let $D(S) = \{S_1, S_2, ..., S_m\}$ denotes the sensitive attributes S of a domain, and let weight (S_i) be the weight of S_i. Then,

$$\text{Weight}(s_i) = \frac{i-1}{m-1}; \quad 1 < i < m$$
$$\text{Weight}(s_k) = 1$$

(1)

Let there be two groups in which all the diseases are classified (as shown in Table 5). We take one disease from each group and make a group of them. Here, A = {HIV, Flu}. According to the formula (1), weight (HIV) = 0, and weight (Flu) = 1. So the weight of A is 0 + 1 = 1. This technique is more enhanced than the 'p-sensitive k-anonymity' because it well protects the sensitive attribute

Table 4 (3, 1)-sensitive 4-anonymous microdata

S.No.	Zip code	Age	Nationality	Disease	Weight	Total
1	123**	<40	*	HIV	0	1
2	123**	<40	*	HIV	0	
3	123**	<40	*	Cancer	0	
4	123**	<40	*	Flu	1	
5	4567*	>40	*	Cancer	0	2
6	4567*	>40	*	Flu	1	
7	4567*	>40	*	HIV	0	
8	4567*	>40	*	Indigestion	1	

Table 5 Categories of disease

Category ID	Sensitive values	Sensitivity
One	HIV, Cancer	Top secret
Two	Flu, Indigestion	Non secret

disclosure than 'p-sensitive k-anonymity.' The limitation of this technique is that in the first QI group, we can see that three tuples belong to the same category of diseases (i.e., Cancer and HIV) and these diseases belong to the top secret category of diseases. So, attacker can identify that the victim has HIV or Cancer with the accuracy of 75%.

2.3 Negative Data Publication

In the negative data publication, we replace the original sensitive values by negative values. These negative values are not taken from outside from the database; in this, we replace the values within the table. We use l-diversity technique to make a negative data publishing. There is a new technique named as SvdND Pused to publish the negative data publication. In l-diversity, we know that there are $l - 1$ distinct sensitive values, and on these values, we apply negative values. To understand this, we take a table which shows a 2-diverse data (Table 6).

Now, we apply the SvdNPD on this table. We transfer the sensitive values into another one according to the sensitive attribute distribution. This is shown in Table 7.

The difference between the 2-diverse table and SvdNDP table is that the later replaces the sensitive values in the l-diverse table with negative values. So here we change the respective values of tuples. So by using this, the attribute disclosure attack is prevented, because if we know the identity of a person, we cannot know the exact value of sensitive attribute. The probability of finding the exact value is $1/(n - 1)$, where n is the number of negative values. But this technique does not give better result when there is a large number of quasi-identifiers; in that condition, it gives less privacy than the l-diversity.

Table 6 2-diverse table

Sex	Zip	Age	Disease
F/M	456**	2*	Cancer
F/M	456**	2*	Dyspepsia
F/M	456**	2*	Gastritis
F/M	456**	2*	Flu

Table 7 SvdNDP table

Sex	Zip	Age	Disease
F/M	456**	2*	Gastritis
F/M	456**	2*	Cancer
F/M	456**	2*	Dyspepsia
F/M	456**	2*	Flu

Table 8 Comparison of various techniques

Privacy techniques	Attacks models				Data utility loss
	Attribute disclosure	Skewness attack	Similarity attack	Probabilistic attack	
K-anonymity	✓	✗	✓	✗	No
L-diversity	✗	✓	✓	✗	No
t-closeness	✗	✗	✗	✗	Yes
p-sensitive K-anonymity	✗	✗	✗	✓	No
(p, α)-sensitive K-anonymity	✗	✗	✗	✓	No
SvdNDP	✗	✓	✓	✗	No

Table 9 Sensitive attribute categories

Category	Category values	c_I	c_j
A	HIV, cancer	HIV	Cancer
B	Hepatitis, phthisis	Hepatitis	Phthisis
C	Asthma, obesity	Asthma	Obesity
D	Indigestion, flu	Indigestion	Flu

Table 8 shows the comparison of all the techniques discussed above and type of attacks each technique can prevent. In this table, '✗' shows that the attack is prevented by that technique and '✓' sign shows that the technique does not prevent this attack. In the next section, we will discuss our proposed methodology which aims to provide some more privacy than all these techniques discussed above (Table 9).

3 Enhanced Method for Privacy-Preserving Using K-Anonymity

In the previous section, we discussed some enhanced techniques which ensure more privacy than the k-anonymity, but they have some limitations. So, to improve the efficiency of the published data, to maximize the volume of published data, and to decrease the risk of similarity attack, a new algorithm is proposed named as (P, U)-sensitive k-anonymity. In this proposed method, discriminated union 'U' of sensitive values is taken which ensures that in each equivalence class there are distinct sensitive values.

3.1   (P, U)-Sensitive K-anonymity

The released dataset must satisfy the k-anonymity for each equivalence group, and there are at least $P = K$ distinct sensitive values with discriminated union U over P. Before applying this algorithm, we distribute the microdata in two subtables. The data which contains the sensitive values in the sensitive attribute column is moved in a new table known as ST table. Now, there are two tables. One table contains only the sensitive attributes known as ST table, and another contains only the non-sensitive attributes known as NST table. The identification of sensitive and non-sensitive attributes is based on the sensitivity of that disease, as shown in Table 5. To maintain the privacy, some distortions are added to the ST table and some portion of non-sensitive tuples from the NST table is added to the ST table. Hence, the size of ST table is increased. Now, the NST table will be released directly because it does not contain any sensitive information and it does not disclose any confidential information regarding individuals. The algorithm is applied to the ST table which contains sensitive information. Now, the sensitive attributes in the ST table will be divided into different categories to ensure the privacy and to prevent confidentiality. This helps us to prevent from similarity, background, and probabilistic attack. Figure 1 shows the flow diagram of the (P, U)-sensitive k-anonymity.

Consider an example scenario. Suppose there are eight diseases in the ST table, then they are separated according to the level of confidentiality of the diseases. Suppose D = {Cancer, HIV, Hepatitis, Phthisis, Asthma, Obesity, Indigestion, and Flu}. Now, they are classified into four categories from A to D, as shown in Table 7.

Table 10 shows the original dataset. The dataset is separated into ST table and NST table.

The sensitive tuples are separated from the dataset and stored in the ST table as shown in Table 11. The NST table is published without any modification because it does not contain any confidential values as shown in Table 12.

We have to improve our ST table because it contains only sensitive values, which can lead to homogeneity attack. This can be done by adding some tuples from the NST table. For this, we added exactly double number of tuples as the number of tuples in the ST table. For example, in ST table, there are only four tuples so we add double of four, i.e., 8 tuples from NST table. Table 13 shows the final input table for the proposed methodology. Now, we will apply '(p, U)-sensitive k-anonymity' method on our input dataset, as shown in Table 13. In this technique, there are two output tables. One is the NST table which is released without any modification and another is the output table which is gathered by applying this technique on the final input table. Now, we talk about 'U,' i.e., the discriminated union. First of all, we have to know about the discriminated category.

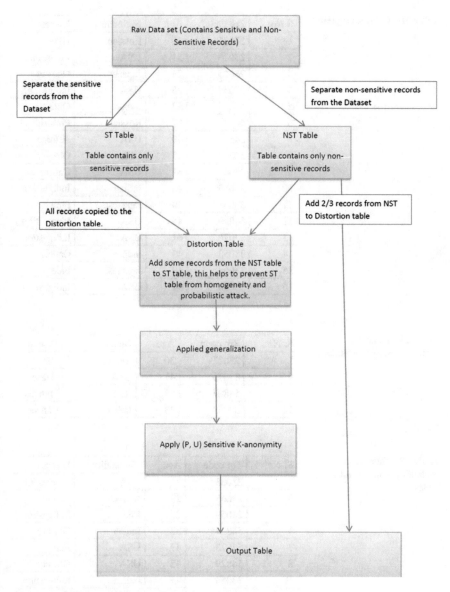

Fig. 1 Flow diagram of the (*P*, *U*)-sensitive *K*-anonymity

3.1.1 Discriminated Category

We can say two categories are discriminated, if they possess no element in common. Suppose A and B are two categories, then they can be called as discriminated categories if they have all distinct values.

Table 10 Dataset original

S.No.	Zip code	Age	Nationality	Disease
1	23021	30	Russian	HIV
2	23023	31	USA	Indigestion
3	23065	23	Japanese	Obesity
4	23059	25	USA	Indigestion
5	54824	52	Indian	Obesity
6	54827	57	Russian	Phthisis
7	54828	49	USA	Asthma
8	54829	45	USA	Obesity
9	33067	41	USA	Cancer
10	33053	43	Indian	Indigestion
11	33068	40	Japanese	Flue
12	54829	45	USA	Flue
13	33076	41	USA	Indigestion
14	33053	43	Indian	Cancer
15	33068	40	Japanese	Flue
16	33068	44	USA	Indigestion

Table 11 Table ST containing sensitive tuples

S. No.	Zip code	Age	Nationality	Disease
1	23021	30	Russian	HIV
9	33076	41	USA	Cancer
6	54827	57	Russian	Phthisis
14	33053	43	Indian	Cancer

Table 12 Table NST containing non-sensitive tuples

S. No.	Zip code	Age	Nationality	Disease
2	23023	31	USA	Indigestion
3	23065	23	Japanese	Obesity
4	23059	25	USA	Indigestion
5	54824	52	Indian	Obesity
7	54828	49	USA	Flue
8	54829	45	USA	Obesity
10	33053	43	Indian	Indigestion
11	33068	40	Japanese	Flue
12	54829	45	USA	Flue
13	33076	41	USA	Indigestion
15	33068	40	Japanese	Flue
16	33068	44	USA	Indigestion

Table 13 Adding some tuples from NST to ST table

S. No.	Zip code	Age	Nationality	Disease
1	23021	30	Russian	HIV
9	33076	41	USA	Cancer
6	54827	57	Russian	Phthisis
14	33053	43	Indian	Cancer
2	23023	31	USA	Indigestion
7	54828	49	USA	Flue
8	54829	45	USA	Obesity
10	33053	43	Indian	Indigestion
11	33068	40	Japanese	Flue
12	54829	45	USA	Flue
3	23065	23	Japanese	Obesity

For example,

$$A = \{Cancer, HIV\}, \quad B = \{Phthisis, Hepatitis\}.$$

3.1.2 Discriminated Union

The discriminated union is the union of sensitive values. It is defined as the set of values which belongs to distinct categories such that no value in the union is overlapping within the set.

The proposed technique says that in any QI group, there is no value of any sensitive category which is repeated. For example, Cancer and HIV belong to one set; if HIV is present in the disease column of QI group, then Cancer is not present in same QI group.

Figure 2 shows the algorithm of proposed technique. Table 14 shows the result of applying proposed technique on Table 13. It contains the sensitive tuples as well as non-sensitive tuples, but it does not contain the full dataset. Table 14 shows that in any QI group, for the sensitive attribute (i.e., disease) values, no two values exist which are from same category. So basically, we can protect our database from homogeneity attack, background knowledge attack, and also similarity attack.

4 Results and Analysis

This section presents the results that we found by applying the proposed technique. The dataset used has been downloaded from the UCI machine learning warehouse [13]. We took the adult dataset for testing our technique. This dataset is publically available, and it has more than 48,000 records. After eliminating the null values and the tuples with unknown values, the dataset is left with 30,000 records.

Step 1:	Scan and make a data set D for sensitive tuples.
Step 2:	Create another table ST and move the sensitive tuples in to that table ST.
Step 3:	Make the NST table from dataset D, import some of tuples in the ST table from NST table, Such that NST = 2ST.
Step 4:	Generalization is applied on the ST table, to make all the tuples equal.
Step 5:	The set G consists of generalized tuples and $U \leftarrow \{ G \}$; $0 \leftarrow \emptyset$.
Step 6:	Repeat
Step 7:	$U \leftarrow \emptyset$
Step 8:	For all $G \in U$ do
Step 9:	Specialize all the tuples in G and go through the generalization hierarchy to satisfying the condition of (P, U) sensitive K-anonymity.
Step 10:	If a node does not satisfy (P, U) sensitive k-anonymity, move it back and un specialized it.
Step 11:	If parent node G does not satisfy the condition, then unspecialized some child node and move upward in the hierarchy to parent so parent full fill the condition of (P, U) sensitive K-anonymity.
Step 12:	End of If structure
Step 13:	For all non-empty branches R of G, do $U' \leftarrow U' \cup \{R\}$
Step 14:	$U \leftarrow U'$
Step 15:	If G is non-empty set then $0 \leftarrow 0 \cup \{G\}$
Step 16:	End of for loop
Step 17:	Until $U = \emptyset$

Fig. 2 Algorithm of the proposed method

Table 14 Dataset with discriminated union 12 of sensitive values $p = k = 2$

S. No.	Zip code	Age	Nationality	Disease
1	230**	3*	*	HIV
2	230**	3*	*	Indigestion
3	230**	3*	*	Obesity
9	330**	4*	*	Cancer
10	330**	4*	*	Indigestion
6	548**	>60	*	Phthisis
7	548**	>60	*	Flue
8	548**	>60	*	Obesity
14	330**	4*	*	Cancer
11	330**	4*	*	Flue
5	5482*	<40	*	Obesity
12	5482*	<40	*	Flue

Now, we see the performance of this technique on the adult dataset. We measure the performance of this technique on three parameters: one is similarity attack, second is data utility or distortion ratio, and the last one is time consumption.

Table 15 Performance of similarity attack

Technique	Parameters	Total QI groups generated	Similarity attack QI groups	Percentage
(P, α)-sensitive K-anonymity	$K = 4$, $P = 2, \alpha = 2$	30 groups	7 groups	7/30 = 23%
(p, U)-sensitive k-anonymity	$K = 4, p = 4$	46 groups	0 groups	0/46 = 0%

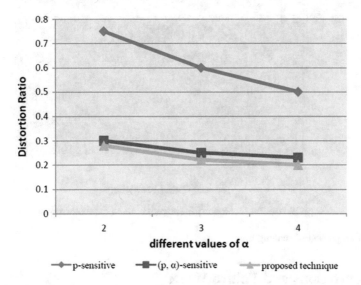

Fig. 3 Graph of data utility or distortion ratio

The performance regarding similarity attack is shown in Table 15. Figures 3 and 4 show the data utility and time-consuming performance of the proposed technique.

In this table, we can see that the previous technique ((P, α)-sensitive K-anonymity) makes 30 QI groups. Out of them, 7 QI groups suffer from homogeneity attack. So, the percentage of that attack is 23%. Our proposed technique makes 46 QI groups, and no group is suffered from similarity attack. Hence, the table is free from similarity attack.

From Figs. 3 and 4, it is clear that this technique gives better result in terms of data utility as well as running time consumption.

Thus, it is a more enhanced technique which gives better protection from all types of attack. Simultaneously, the data utility is also maintained because we applied k-anonymity. There are two tables which are published, one is the anonymized table and second is the NST table which is published without any privacy, thus maintaining data utility.

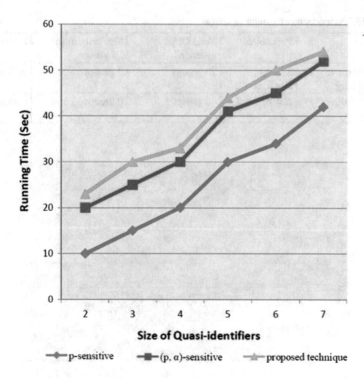

Fig. 4 Comparison of running time

5 Conclusions and Future Work

In this paper, we study about the privacy-preserving data publishing. There are
three main techniques 'k-anonymity,' 'l-diversity,' and 't-closeness' used to preserve the data. These techniques have some limitations which lead to various
attacks such as homogeneity attack, similarity attack, and the rate of data utility
loss. So to overcome these attacks, we proposed a technique named as '(P, U)-
sensitive K-anonymity.' This technique was able to overcome the attacks we discussed in the paper. In the previous technique (p, α)-sensitive k-anonymity, we see
that there is a chance of similarity attack and probabilistic attack, but our proposed
technique prevents the dataset from similarity attack with 100% accuracy. It also
overcomes homogeneity attack and probabilistic attack. So it is an enhanced
technique than the previous technique. More than 50% of data is published without
any modification, so the data utility is also maintained. But this does not work
efficiently for very large dataset, so in future we have to make this technique more
reliable so that it can also be used for very large datasets.

References

1. L. Wu, W. Luo, D. Zhao, SvdNPD: a negative data publication method based on the sensitive value distribution, in *2015 International Workshop on Artificial Immune Systems (AIS)* (IEEE, 2015)
2. L. Sweeney, *k*-anonymity: a model for protecting privacy. Int. J. Uncertain. Fuzziness Knowl. Based Syst. **10**(5), 557–570 (2002)
3. A. Machanavajjhala, D. Kifer, J. Gehrke, M. Venkitasubramaniam, *L-Diversity: Privacy Beyond k-Anonymity*. ACM Transactions on Knowledge Discovery from Data (TKDD) (2007)
4. L. Sweeney, Uniqueness of simple demographics in the U.S. population. Technical report, Carnegie Mellon University (2000)
5. N. Li, T. Li, S. Venkatasubramanian, *t*-closeness: privacy beyond k-anonymity and l-diversity, in *IEEE 23rd International Conference on Data Engineering (ICDE 2007)* (IEEE, 2007)
6. C. Clifton et al., Privacy-Preserving Data Integration and Sharing, in *Proceedings of the 9th ACM SIGMOD Workshop on Research Issues in Data Mining and Knowledge Discovery* (ACM, 2004)
7. T. Li, N. Li, On the tradeoff between privacy and utility in data publishing, in *Proceedings of the 15th ACM SIGKDD International Conference on Knowledge Discovery and Data Mining* (ACM, 2009)
8. B. Fung, K. Wang, P. Yu, Top-down specialization for information and privacy preservation, in *Proceedings of the 21st International Conference on Data Engineering (ICDE05)*, Tokyo, Japan (2005)
9. K. LeFevre, D. DeWitt, R. Ramakrishnan, Incognito: efficient full-domain k-anonymity, in *ACM SIGMOD International Conference on Management of Data* (2005)
10. L. Sweeney, Achieving *k*-anonymity privacy protection using generalization and suppression. Int. J. Uncertainty Fuzziness Knowl. Based Syst. **10**(5)
11. X. Sun, H. Wang, J. Li, On the complexity of restricted-anonymity problem, in *Accepted by the 10th Asia Pacific WebConference (APWEB 2008)*, Shenyang, China (2008)
12. T.M. Truta, A. Campan, P. Meyer, Generating micro data with *P*-sensitive *k*-anonymity property. SDM, (2007) pp. 124–141
13. X. Sun, H. Wang, L. Sun, Extended *K*-anonymity models against attribute disclosure. in *3rd International Conference on Network and System Security*, 2009. NSS'09 (IEEE, 2009)

Computational Algorithmic Procedure of Integrated Mixture Distribution Inventory Model with Errors in Inspection

R. Uthayakumar and M. Ganesh Kumar

Abstract When the lead time demand of different customers is not identical, we could not able to use a single distribution to describe the demand of the lead time. Hence, in this paper, we implement a mixture distribution inventory model with variable lead time, and integrated production for two echelon is also investigated where buyer follows a continuous review policy (Q, r). We assumed the production process as imperfect one, so the buyer inspects the items after arrival of each lot. During the inspection process, the items may be misclassified by the buyer, i.e., the buyer may classify good items as defective one or vice versa. Moreover, the lead time crashing cost is a function of L, and it is a negative exponential function. Our aim is to minimize the integrated total expected cost which is a nonlinear function. So we present an algorithm to find the optimal solutions. Sensitive analysis is performed to show how the different values of the parameter affect the entire system.

Keywords Mixture of inventory · Inspection process · Lead time crashing cost Integrated approach

1 Introduction

Amid the lead time, demands of various buyers are not same, and the probability distribution of demand for each buyer can be enough approximated by a probability distribution. The overall distribution during demand is then a mixture. Along these lines, we cannot use just one distribution (such as [1–4] with a normal distribution) to depict the request of lead time. Therefore, in this article, we implement the

R. Uthayakumar · M. Ganesh Kumar (✉)
Department of Mathematics, The Gandhigram Rural Institute—Deemed University,
Gandhigram, Dindigul 624302, Tamil Nadu, India
e-mail: ganeshkumarnov23@gmail.com

R. Uthayakumar
e-mail: uthayagri@gmail.com

© Springer Nature Singapore Pte Ltd. 2018   77
B. Panda et al. (eds.), *Innovations in Computational Intelligence*, Studies in
Computational Intelligence 713, https://doi.org/10.1007/978-981-10-4555-4_6

mixtures of normal distribution as in [5], i.e., the lead time demand is stochastic (mixture of normal) and whose cumulative distribution function is $F_* = pF_1 + (1 - p)F_2$, where F_1 and F_2 are cumulative distribution functions and $0 \leq p \leq 1$. References [6–8] are employed the mixture of inventory model. In addition, we assumed that shortages are allowed. Moreover, the total shortages are considered to be a mixture of back orders and lost sales during the stock-out period. Many of the researchers in the inventory problem focused their attention into the integrated strategy. For production network administration, building up long haul key organizations between the purchaser and seller is worthwhile for the two gatherings in regard to costs, and in this manner benefits, since both sides can accomplish enhanced advantages, collaborate and impart data to each other. Several researchers [2, 3, 9, 10] have demonstrated that the purchaser and merchant can accomplish their own insignificant aggregate cost, or increase their shared advantage through key collaboration with each other.

Quality inspection is an essential tool in the industrial management. In particular, pharmaceutical, aircraft, food production, and other firms the screening process places an important role. Initially, Bennett et al. [11] investigated the effects of inspection errors on a cost-based single examining arrangement. Many researchers addressed the inspection errors on the performance measures of a complete repeat inspection plan, and they developed EOQ/EPQ inventory model with imperfect production and inspection errors. Recently, Dey and Giri [2] and Hsien-Jen [4] assumed that there is a fixed percentage of defective items in every lot-size Q. But they assumed that inspection process is 100% error free. Khan et al. [12] developed a mathematical model to find an optimal vendor–buyer inventory policy. They considered quality inspection errors at the buyer's end and learning in production at vendor's end with a goal to minimize the joint annual cost incurred by the supply chain.

In the literature, lead time is seen as an endorsed consistent or a stochastic variable, which along these lines, is not subject to control for managing inventory problems. In fact, Tersine [13] thought that the lead time usually consists of the n mutually different components such as set up time order preparation, order transit, supplier lead time, and delivery time. So, the crashing cost function is a piecewise linear function. The lead time can be shortened by adding some crashing cost. Similar lead time reduction is employed by Ben-Daya and Raouf [14]. But here the crashing cost function is negative exponential. We adopt a negative exponential crashing cost function for this model to reduce the lead time.

Reference [15] is first who considered that among the shortages, a fraction of shortage is back ordered and the remaining fraction of shortages is lost. Ouyang et al. [16] considered an inventory model with a mixture of back orders and lost sales to generalize [14] model, where the back order rate is constant. However, Ouyang et al. [16] investigated a mixture inventory problem with variable lead time and variable backlogging rate. They observe that many consumable items of famous brands or fashionable goods such as certain hi-fi equipment, brand gum shoes (or leather shoes), clothes, and cosmetics may lead to a situation in which customers prefer their demands to be back ordered when stock out occur. Obviously, if the shortages are accumulated to a degree that overlapping the waiting patience of customers, some consumers may refuse the backlogging case. In addition, Annadurai and Uthayakumar [17] developed ordering policy for periodic review inventory model with controllable lead time by reducing lost sales rate.

The remainder of this paper is organized as follows. In Sect. 2, we provide the notations and assumptions. The mathematical model of mixture normal distribution model for single item with quality inspection error is employed in Sect. 3. Solution procedure is developed in Sect. 4. We present numerical examples and sensitivity analysis in Sect. 5. And finally in Sect. 6, the conclusion of this study is summarized.

2 Notations and Assumptions

To develop the proposed model, we adopt the following notation and assumptions which are similar to [8, 9].

2.1 Notations

The following notations are used in this paper

n	Number of shipments, a decision variable
Q	Economic order quantity, a decision variable
A	Ordering cost per order at the buyer side
h_b	Holding cost per unit at the buyer side
h_v	Holding cost per unit at the vendor side and $h_v < h_b$
B	Vendor's setup cost per set up
T_r	Buyer transportation cost per delivery
L	Length of the lead time, a decision variable
D	Demand rate
P	Production rate and $P > D$
v	Buyers unit variable cost for order receiving and handling ($/unit)
C_v	Production cost per unit time at the vendor side
s	Buyer's screening cost per unit
x	Screening rate
y	Defective percentage among the lot-size Q, and it is a random variable
y_c	Defective percentage among the lot-size Q due to the buyer's observation $y_c = (1 - y)y' + y(1 - y'')$
y'	Probability of 1st type error (classifying a usable product as defective)
y''	Probability of 2nd type error (classifying a defective product as usable)
$f(y)$	Probability density function of y
$f(y')$	Probability density function of y'

$f(y'')$ Probability density function of y''
θ False acceptance cost rate per unit
θ_0 False rejection cost rate per unit
T Time duration between successive replenishment
π Fixed penalty cost per unit short
π_0 Penalty cost for the lost demand per unit short
β Fraction of the shortages back ordered into the next shipment
r Reorder level
X The lead time demand with the mixture of distributions as in [5]
a^+ Maximum value of a and 0
a^- Maximum value of $-a$ and 0
$\mathbb{E}(\cdot)$ Mathematical expectation

2.2 Assumptions

1. The integrated inventory system for single item is considered as in [18]. According to Goyal, the sum of the total cost of the individuals is greater than the total cost of the integrated approach.
2. Since the production rate is greater than the demand rate, even though the shortages occur because there is a lead time in every shipment. We assumed the demand is stochastic, so the consumption during the lead time may excess the on-hand inventory level.
3. The demand during out of stock period (shortages) is partially backlogged to the next shipment, i.e., the fraction β of shortage is back ordered in the next cycle, and remaining shortages are lost.
4. We assume that the demand during the lead time X has the mixture of cumulative distribution function F_* with finite mean $\mu_* L$ and standard deviation $\sigma_* \sqrt{L}$, where $F_* = pF_1 + (1-p)F_2$, F_1 has a finite mean $\mu_1 L$, standard deviation $\sigma\sqrt{L}$, F_2 has finite mean $\mu_2 L$, standard deviation $\sigma\sqrt{L}$, $\mu_1 - \mu_2 = \epsilon\sigma/\sqrt{L}$, $\epsilon \in \mathbb{R}$, and $0 \le p \le 1$.
5. The buyer follows continuous review policy (Q, r). The reorder point $r =$ expected consumption during lead time + safety stock (ss), and ss $= kx$ (standard deviation of lead time demand). That is, $r = \mu_* L + k\sigma_* \sqrt{L}$, where $\mu_* = p\mu_1 + (1-p)\mu_2$, $\sigma_* = \sigma\sqrt{1 + p(1-p)\epsilon^2}$, $\mu_1 = \mu_* + (1-p)\epsilon\sigma/\sqrt{L}$, $\mu_1 = \mu_* + (1-p)\epsilon\sigma/\sqrt{L}$ and k is the safety factor.
6. The lead time crashing cost is a function of L, and it is given by

$$R(L) = \eta_1 e^{-\eta_2 L}, \tag{1}$$

where η_1, η_2 are positive constants. $R(L)$ also called negative exponential crashing cost, and it is a decreasing function. The basic idea of choosing this type of functions is whenever the buyer wants to reduce the lead time, he has to pay that much money.

3   Mathematical Model

Since we assumed that the demand of the lead time X follows mixture of normal distribution, its probability density function is

$$
\begin{aligned}
f(x) = p\, &\frac{1}{\sqrt{2\pi}\sigma\sqrt{L}}\exp\left[-\frac{1}{2}\frac{(x-\mu_1 L)^2}{\sigma^2 L}\right] \\
&+ (1-p)\frac{1}{\sqrt{2\pi}\sigma\sqrt{L}}\exp\left[-\frac{1}{2}\frac{(x-\mu_2 L)^2}{\sigma^2 L}\right], \quad 0 \le p \le 1,
\end{aligned}
\tag{2}
$$

and the reorder point $r = \mu_* L + k\sigma_* \sqrt{L}$, where k, μ_*, and σ_* are defined above. The expected shortage at the end of the cycle is

$$
\begin{aligned}
\mathbb{E}(X-r)^+ &= \int_r^\infty (x-r)dF_*(x), \\
&= \sigma_i\sqrt{L}\Psi(r_1, r_2, p),
\end{aligned}
\tag{3}
$$

where $\Psi(r_1, r_2, p) = p[\phi(r_1) - r_1(1 - F(r_1))] + (1-p)[\phi(r_2) - r_1(1 - F(r_2))]$, $r_1 = k\sqrt{1+\epsilon^2 p(1-p)} - \epsilon_i(1-p)$ and $r_2 = k\sqrt{1+\epsilon^2 p(1-p)} + \epsilon p$, ϕ and F denote the standard normal probability density function and cumulative distribution function, respectively. The safety factor k satisfies

$$
P(X > r) = 1 - pF(r_1) - (1-p)F(r_2) = q,
\tag{4}
$$

where q represents the allowable stock-out probability during L.

3.1   Buyer's Perspective

The fraction β of shortages is back ordered, and the remaining $(1-\beta)\mathbb{E}(X-r)^+$ shortages are lost. So the cost of lost sale and back ordering for all n shipments is

$$n[\pi + \pi_0(1 - \beta)]\mathbb{E}(X - r)^+. \tag{5}$$

and sum of the buyer's transportation cost, screening cost, lead time crashing cost, lost sale and back ordering cost, unit cost for order receiving and handling for all n shipments is

$$A + nQx + n(T_r + vQ) + nR(L) + n[\pi + \pi_0(1 - \beta)]\mathbb{E}(X - r)^+. \tag{6}$$

The net inventory level before the arrival of an order is

$$\mathbb{E}(X - r)^- I_{(0 < X < r)} - \beta \mathbb{E}(X - r)^+,$$
$$= \sigma\sqrt{L}\left\{ p\left[r_1 F\left(\frac{\mu_* \sqrt{L}}{\sigma} + (1 - p)\epsilon\right) - \phi\left(\frac{\mu_* \sqrt{L}}{\sigma}(1 - p)\epsilon\right)\right] \right.$$
$$\left. + (1 - p)\left[r_2 F\left(\frac{\mu_* \sqrt{L}}{\sigma} - p\epsilon\right) - \phi\left(\frac{\mu_* \sqrt{L}}{\sigma} - p\epsilon\right)\right] \right\} + (1 - \beta)\mathbb{E}(X - r)^+,$$

where $I_{(0 < X < r)} = \begin{cases} 1, & 0 < x < r, \\ 0, & \text{otherwise,} \end{cases}$

$$\tag{7}$$

and the net inventory level at the beginning of the cycle is

$$Q + \sigma\sqrt{L}\left\{ p\left[r_1 F\left(\frac{\mu_* \sqrt{L}}{\sigma} + (1 - p)\epsilon\right) - \phi\left(\frac{\mu_* \sqrt{L}}{\sigma}(1 - p)\epsilon\right)\right] \right.$$
$$\left. + (1 - p)\left[r_2 F\left(\frac{\mu_* \sqrt{L}}{\sigma} - p\epsilon\right) - \phi\left(\frac{\mu_* \sqrt{L}}{\sigma} - p\epsilon\right)\right] \right\} + (1 - \beta)\mathbb{E}(X - r)^+.$$

$$\tag{8}$$

The buyer screens all Q items, and he separates defective and non-defective items at the finite screening rate x. Therefore, expected inventory level for all n cycles is

$$n\left[\frac{Q(1 - \mathbb{E}[y])\mathbb{E}(T)}{2} + \frac{Q}{x}Q\mathbb{E}(y)\right], \quad \text{where} \quad \mathbb{E}(T) = \frac{Q(1 - \mathbb{E}[y])}{D}, \tag{9}$$

and the inventory level at any time is given by

$$\frac{n}{n\mathbb{E}(T)}\left[\frac{Q(1 - \mathbb{E}[y])\mathbb{E}(T)}{2} + \frac{Q}{x}Q\mathbb{E}(y)\right], \tag{10}$$

the above equation can rearranged in the form of

$$\frac{Q(1 - \mathbb{E}[y])}{2} + \frac{QD\mathbb{E}(y)}{x(1 - \mathbb{E}[y])}. \tag{11}$$

The buyer's expected total cost per unit time is given by

$$\begin{aligned}
\text{ETCB} &= \frac{AD}{nQ(1 - \mathbb{E}[y])} + \frac{xD}{1 - \mathbb{E}(y)} + \frac{D[\pi + \pi_0(1 - \beta)]\sigma\sqrt{L}\Psi(r_1, r_2, p)}{Q(1 - \mathbb{E}[y])} \\
&+ \frac{(T_r + vQ)D}{Q(1 - \mathbb{E}[y])} + \frac{R(L)D}{Q(1 - \mathbb{E}[y])} + h_b\left[\frac{Q(1 - \mathbb{E}[y])}{2} + \frac{QD\mathbb{E}(y)}{x(1 - \mathbb{E}[y])}\right] \\
&+ h_b\left\{\sigma\sqrt{L}\left\{p\left[r_1 F\left(\frac{\mu_*\sqrt{L}}{\sigma} + (1 - p)\epsilon\right) - \phi\left(\frac{\mu_*\sqrt{L}}{\sigma}(1 - p)\epsilon\right)\right]\right.\right. \\
&+ (1 - p)\left[r_2 F\left(\frac{\mu_*\sqrt{L}}{\sigma} - p\epsilon\right) - \phi\left(\frac{\mu_*\sqrt{L}}{\sigma} - p\epsilon\right)\right]\right\} \\
&+ (1 - \beta)\sigma\sqrt{L}\Psi(r_1, r_2, p)\}.
\end{aligned} \tag{12}$$

3.2 Vendor's Perspective

Finding the vendor inventory level is similar to [19]. The vendor total cost which is the sum of the setup cost, holding cost, and production cost is given below

$$B + h_v\frac{nQ^2}{2D}\left\{(n - 1) - (n - 2)\frac{D}{P}\right\} + \frac{nQC_v}{P}. \tag{13}$$

Therefore, the vendor's total cost per unit time is

$$\text{ETCV} = \frac{BD}{nQ(1 - \mathbb{E}[y])} + h_v\frac{Q}{2(1 - \mathbb{E}[y])}\left\{(n - 1) - (n - 2)\frac{D}{P}\right\} + \frac{C_vD}{P(1 - \mathbb{E}[y])}. \tag{14}$$

3.3 Integrated Approach

In the integrated approach, the total cost is sum of the total costs of vendor and buyer and takes the total cost function as objective function. Our aim is to minimize this cost function with respect to the decision variables. Hence, the integrated expected total cost is

$$TC = ETCB + ETCV,$$

$$= \left[\frac{A+B}{n} + \eta_1 e^{-\eta_2 L} + T_R\right]\frac{D}{Q(1-\mathbb{E}[y])} + \left(v + \frac{C_v}{P} + x\right)\frac{D}{1-\mathbb{E}(y)}$$

$$+ h_b\left[\frac{Q(1-\mathbb{E}[y])}{2} + \frac{QD\mathbb{E}(y)}{x(1-\mathbb{E}[y])}\right] + h_v\frac{Q}{2(1-\mathbb{E}[y])}\left\{(n-1)-(n-2)\frac{D}{P}\right\}$$

$$+ h_b\left\{\sigma\sqrt{L}\left\{p\left[r_1 F\left(\frac{\mu_*\sqrt{L}}{\sigma} + (1-p)\epsilon\right) - \phi\left(\frac{\mu_*\sqrt{L}}{\sigma}(1-p)\epsilon\right)\right]\right.\right.$$

$$+ (1-p)\left[r_2 F\left(\frac{\mu_*\sqrt{L}}{\sigma} - p\epsilon\right) - \phi\left(\frac{\mu_*\sqrt{L}}{\sigma} - p\epsilon\right)\right]\right\}$$

$$+ (1-\beta)\sigma\sqrt{L}\Psi(r_1,r_2,p)\} + \frac{D[\pi+\pi_0(1-\beta)]\sigma\sqrt{L}\Psi(r_1,r_2,p)}{Q(1-\mathbb{E}[y])}.$$

$$(15)$$

3.4 Inspection Errors

During the inspection process, buyer may classify the good items as a useless one or vice versa. That is, the buyer classify some usable items as defectives, i.e., $(1-y)y'$, while some defective units as usable items, i.e., yy''. Thus, the percentage of non-usable units as professed by the buyer is

$$y_c = (1-y)y' + y(1-y''),$$
$$\text{thus } \mathbb{E}[y_c] = (1-\mathbb{E}[y])\mathbb{E}[y'] + \mathbb{E}[y](1-\mathbb{E}[y'']).$$

$$(16)$$

So the time interval between successive shipments would now be

$$\mathbb{E}[T] = \frac{(1-\mathbb{E}[y_c])Q}{D} = \frac{[1-(1-\mathbb{E}[y])\mathbb{E}[y'] - \mathbb{E}[y](1-\mathbb{E}[y''])]Q}{D}.$$

$$(17)$$

Since we assume that θ and θ_0 are the unit fraction cost of false acceptance and false rejection, respectively, where θ and θ_0 are lie in the unit interval $[0, 1]$. The safety purpose is the 1st preference, and the cost of false acceptance should be greater than that of false rejection. According to this fact, the value of θ is chosen to be greater than θ_0. Therefore, integrated expected total cost in (15) will become

$$
\begin{aligned}
\text{ETC}(Q, L, n) = {} & \left[\frac{A+B}{n} + \eta_1 e^{-\eta_2 L} + T_R\right]\frac{D}{Q(1 - \mathbb{E}[y_c])} + \left(v + \frac{C_v}{P} + x\right)\frac{D}{1 - \mathbb{E}[y_c]} \\
& + h_b\left[\frac{Q(1 - \mathbb{E}[y_c])}{2} + \frac{QD\mathbb{E}(y_c)}{x(1 - \mathbb{E}(y_c))}\right] + h_v\frac{Q}{2(1 - \mathbb{E}(y_c))}\left\{(n-1) - (n-2)\frac{D}{P}\right\} \\
& + h_b\left\{\sigma\sqrt{L}\left\{p\left[r_1 F\left(\frac{\mu_*\sqrt{L}}{\sigma} + (1-p)\epsilon\right) - \phi\left(\frac{\mu_*\sqrt{L}}{\sigma}(1-p)\epsilon\right)\right]\right.\right. \\
& + (1-p)\left[r_2 F\left(\frac{\mu_*\sqrt{L}}{\sigma} - p\epsilon\right) - \phi\left(\frac{\mu_*\sqrt{L}}{\sigma} - p\epsilon\right)\right]\right\} \\
& + (1-\beta)\sigma\sqrt{L}\Psi(r_1, r_2, p)\} + \frac{D[\pi + \pi_0(1-\beta)]\sigma\sqrt{L}\Psi(r_1, r_2, p)}{Q(1 - \mathbb{E}[y_c])} \\
& + \frac{\theta Q\mathbb{E}(y)\mathbb{E}(y'')D}{1 - \mathbb{E}(y_c)} + \frac{\theta_0 Q(1 - \mathbb{E}[y])\mathbb{E}[y']D}{1 - \mathbb{E}(y_c)}.
\end{aligned}
\tag{18}
$$

The cost function is an objective function of our model. We have to minimize this function with respect to the decision variables. They are economic order quantity Q, lead time L, and number of deliveries n.

4 Solution Procedure

The problem formulated in Sect. 3 appears as a nonlinear programming problem. To solve this type of nonlinear problem, we use the similar method of most of the literature dealing with nonlinear problem. First of all, we want to know the convexity and concavity of the total cost function with respect to the decision variables. For fixed Q and L, ETC is convex in n, which indicates that there is a $n = n^*$ satisfying the following relation

$$
\text{ETC}(Q, L, n^* - 1) \geq \text{ETC}(Q, L, n^*) \leq \text{ETC}(Q, L, n^* + 1),
$$

$$
\text{because} \quad \frac{\partial}{\partial n}\text{ETC} = -\frac{1}{n^2}(A+B)\frac{D}{Q(1 - \mathbb{E}(y_c))} + h_v\frac{QD}{2P(1 - \mathbb{E}(y_c))}, \tag{19}
$$

$$
\text{and} \quad \frac{\partial^2}{\partial n^2}\text{ETC} = \frac{2(A+B)D}{n^3 Q(1 - \mathbb{E}(y_c))} \quad \text{for all} \quad n > 0.
$$

Now taking the partial derivatives of $\text{ETC}(Q, L, n)$ for Q and L, we obtain

$$
\begin{aligned}
\frac{\partial}{\partial Q}\text{ETC}(Q, L, n) = {} & -\frac{D}{Q^2(1 - \mathbb{E}(y_c))}\left[\frac{A+B}{n} + \eta_1 e^{-\eta_2 L} + T_R\right] \\
& + h_b\left[\frac{1 - \mathbb{E}[y_c]}{2} + \frac{D}{x}\frac{\mathbb{E}(y_c)}{1 - \mathbb{E}(y_c)}\right] - \frac{D[\pi + \pi_0(1-\beta)]}{Q^2(1 - \mathbb{E}[y_c])}\sigma\sqrt{L}\Psi(r_1, r_2, p).
\end{aligned}
\tag{20}
$$

and

$$
\frac{\partial}{\partial L} \text{ETC}(Q, L, n) = -\frac{D}{Q(1 - \mathbb{E}[y_c])} \eta_1 \eta_2 e^{-\eta_2 L} + h_b \frac{(1 - \beta)\sigma \Psi(r_1, r_2, p)}{2\sqrt{L}}
$$
$$
+ \frac{h_b \sigma}{2\sqrt{L}} \left\{ p \left[r_1 F\left(\frac{\mu_* \sqrt{L}}{\sigma} + (1 - p)\epsilon\right) - \phi\left(\frac{\mu_* \sqrt{L}}{\sigma}(1 - p)\epsilon\right) \right] \right.
$$
$$
+ (1 - p) \left[r_2 F\left(\frac{\mu_* \sqrt{L}}{\sigma} - p\epsilon\right) - \phi\left(\frac{\mu_* \sqrt{L}}{\sigma} - p\epsilon\right) \right] \right\}
$$
$$
+ \frac{h_b}{2} \mu_* \left[p\left(r_1 + \frac{\mu_* \sqrt{L}}{\sigma} + (1 - p)\epsilon\right) \phi\left(\frac{\mu_* \sqrt{L}}{\sigma} + (1 - p)\epsilon\right) \right.
$$
$$
+ (1 - p)\left(r_2 + \frac{\mu_* \sqrt{L}}{\sigma} - p\epsilon\right) \phi\left(\frac{\mu_* \sqrt{L}}{\sigma} - p\epsilon\right) \right]
$$
$$
+ \frac{D[\pi + \pi_0(1 - \beta)]}{Q(1 - \mathbb{E}[y_c])} \frac{\sigma \Psi(r_1, r_2, p)}{2\sqrt{L}}.
$$

$$(21)$$

respectively. Setting Eqs. (20) and (21) equal to zero and solving with respect to Q and L, we obtain

$$
Q = \left[\frac{2D\left[\frac{A+B}{n} + \eta_1 e^{-\eta_2 L} + T_r + (\pi + \pi_0(1 - \beta))\sigma\sqrt{L}\Psi(r_1, r_2, p)\right]}{h_b\left[(1 - \mathbb{E}[y_c])^2 + \frac{2D}{x}\mathbb{E}[y_c]\right] + h_v\left[(n - 1) - (n - 2)\frac{D}{P}\right] + G} \right]^{\frac{1}{2}},
$$

$$(22)$$

where $G = 2\theta\mathbb{E}(y)\mathbb{E}(y'')D + 2\theta_0(1 - \mathbb{E}[y])\mathbb{E}[y']D$, and

$$
\frac{D}{Q(1 - \mathbb{E}[y_c])} \eta_1 \eta_2 e^{-\eta_2 L} = h_b \frac{(1 - \beta)\sigma \Psi(r_1, r_2, p)}{2\sqrt{L}} + \frac{D[\pi + \pi_0(1 - \beta)]}{Q(1 - \mathbb{E}[y_c])} \frac{\sigma \Psi(r_1, r_2, p)}{2\sqrt{L}}
$$
$$
+ \frac{h_b \sigma}{2\sqrt{L}} \left\{ p \left[r_1 F\left(\frac{\mu_* \sqrt{L}}{\sigma} + (1 - p)\epsilon\right) - \phi\left(\frac{\mu_* \sqrt{L}}{\sigma}(1 - p)\epsilon\right) \right] \right.
$$
$$
+ (1 - p) \left[r_2 F\left(\frac{\mu_* \sqrt{L}}{\sigma} - p\epsilon\right) - \phi\left(\frac{\mu_* \sqrt{L}}{\sigma} - p\epsilon\right) \right] \right\}
$$
$$
+ \frac{h_b}{2} \mu_* \left[p\left(r_1 + \frac{\mu_* \sqrt{L}}{\sigma} + (1 - p)\epsilon\right) \phi\left(\frac{\mu_* \sqrt{L}}{\sigma} + (1 - p)\epsilon\right) \right.
$$
$$
+ (1 - p)\left(r_2 + \frac{\mu_* \sqrt{L}}{\sigma} - p\epsilon\right) \phi\left(\frac{\mu_* \sqrt{L}}{\sigma} - p\epsilon\right) \right].
$$

$$(23)$$

It is obvious that for any k(or q) and p, we have $\Psi(r_1, r_2, p) > 0$. Hence

$$\frac{\partial^2}{\partial Q^2}\text{ETC} = \frac{2D}{Q^3(1 - \mathbb{E}[y_c])}\left[\frac{A+S}{n} + \eta_1 e^{-\eta_2 L} + T_r + [\pi + \pi_0(1 - \beta)]\sigma L^{\frac{1}{2}}\Psi(r_1, r_2, p)\right] > 0.$$

(24)

In addition, we also obtain

$$\begin{aligned}
\frac{\partial^2}{\partial L^2}\text{ETC} = &-(\pi + \pi_0(1 - \beta))\frac{\sigma\Psi(r_1, r_2, p)D}{4QL^{\frac{3}{2}}} - h_b\frac{(1 - \beta)\sigma\Psi(r_1, r_2, p)}{4L^{\frac{3}{2}}}\\
&- h_b\frac{\sigma}{4L^{\frac{3}{2}}}\left\{p\left[r_1 F\left(\frac{\mu_*\sqrt{L}}{\sigma} + (1 - p)\epsilon\right) - \phi\left(\frac{\mu_*\sqrt{L}}{\sigma} + (1 - p)\epsilon\right)\right]\right.\\
&\left.+ (1 - p)\left[r_2 F\left(\frac{\mu_*\sqrt{L}}{\sigma} - p\epsilon\right) - \phi\left(\frac{\mu_*\sqrt{L}}{\sigma} - p\epsilon\right)\right]\right\}\\
&- h_b\frac{\mu_*}{4L}\left\{p\phi\left(\frac{\mu_*\sqrt{L}}{\sigma} + (1 - p)\epsilon\right)\left[\frac{\mu_*\sqrt{L}}{\sigma}\left(\frac{\mu_*\sqrt{L}}{\sigma} + (1 - p)\epsilon\right)\left(r_1 + \frac{\mu_*\sqrt{L}}{\sigma} + (1 - p)\epsilon\right)\right.\right.\\
&\left.- \left(r_1 + \frac{2\mu_*\sqrt{L}}{\sigma} + (1 - p)\epsilon\right)\right] + (1 - p)\phi\left(\frac{\mu_*\sqrt{L}}{\sigma} - p\epsilon\right)\left[\frac{\mu_*\sqrt{L}}{\sigma}\left(\frac{\mu_*\sqrt{L}}{\sigma} - p\epsilon\right)\right.\\
&\left.\left.\times \left(r_2 + \frac{\mu_*\sqrt{L}}{\sigma} - p\epsilon\right) - \left(r_2 + \frac{2\mu_*\sqrt{L}}{\sigma} - p\epsilon\right)\right]\right\} + \frac{D}{Q(1 - \mathbb{E}[y_c])}\eta_1\eta_2^2 e^{-\eta_2 L}\\
&- \left\{h_b(1 - \beta) + \frac{D[\pi + \pi_0(1 - \beta)]}{Q(1 - \mathbb{E}[y_c])}\right\}\frac{\sigma\Psi(r_1, r_2, p)}{4L\sqrt{L}},
\end{aligned}$$

(25)

and

$$\begin{aligned}
&\frac{\partial^2\text{ETC}}{\partial Q^2} \times \frac{\partial^2\text{ETC}}{\partial L^2} - \left(\frac{\partial^2\text{ETC}}{\partial Q\partial L}\right)^2\\
&= \frac{2D}{Q^3(1 - \mathbb{E}[y_c])}\left[\frac{A+S}{n} + \eta_1 e^{-\eta_2 L} + T_r + [\pi + \pi_0(1 - \beta)]\sigma L^{\frac{1}{2}}\Psi(r_1, r_2, p)\right]\\
&\times \left\{-(\pi + \pi_0(1 - \beta))\frac{\sigma\Psi(r_1, r_2, p)D}{4QL^{\frac{3}{2}}} - h_b\frac{(1 - \beta)\sigma\Psi(r_1, r_2, p)}{4L^{\frac{3}{2}}}\right.\\
&- h_b\frac{\sigma}{4L^{\frac{3}{2}}}\left\{p\left[r_1 F\left(\frac{\mu_*\sqrt{L}}{\sigma} + (1 - p)\epsilon\right) - \phi\left(\frac{\mu_*\sqrt{L}}{\sigma} + (1 - p)\epsilon\right)\right]\right.\\
&\left.+ (1 - p)\left[r_2 F\left(\frac{\mu_*\sqrt{L}}{\sigma} - p\epsilon\right) - \phi\left(\frac{\mu_*\sqrt{L}}{\sigma} - p\epsilon\right)\right]\right\}\\
&- h_b\frac{\mu_*}{4L}\left\{p\phi\left(\frac{\mu_*\sqrt{L}}{\sigma} + (1 - p)\epsilon\right)\left[\frac{\mu_*\sqrt{L}}{\sigma}\left(\frac{\mu_*\sqrt{L}}{\sigma} + (1 - p)\epsilon\right)\left(r_1 + \frac{\mu_*\sqrt{L}}{\sigma} + (1 - p)\epsilon\right)\right.\right.\\
&\left.- \left(r_1 + \frac{2\mu_*\sqrt{L}}{\sigma} + (1 - p)\epsilon\right)\right] + (1 - p)\phi\left(\frac{\mu_*\sqrt{L}}{\sigma} - p\epsilon\right)\left[\frac{\mu_*\sqrt{L}}{\sigma}\left(\frac{\mu_*\sqrt{L}}{\sigma} - p\epsilon\right)\right.\\
&\left.\left.\times \left(r_2 + \frac{\mu_*\sqrt{L}}{\sigma} - p\epsilon\right) - \left(r_2 + \frac{2\mu_*\sqrt{L}}{\sigma} - p\epsilon\right)\right]\right\} + \frac{D}{Q(1 - \mathbb{E}[y_c])}\eta_1\eta_2^2 e^{-\eta_2 L}\\
&\left.- \left\{h_b(1 - \beta) + \frac{D[\pi + \pi_0(1 - \beta)]}{Q(1 - \mathbb{E}[y_c])}\right\}\frac{\sigma\Psi(r_1, r_2, p)}{4L\sqrt{L}}\right\}\\
&- \left[\frac{D}{Q(1 - \mathbb{E}[y_c])}\left(\eta_1\eta_2 e^{-\eta_2 L} - [\pi + \pi_0(1 - \beta)]\frac{\sigma\Psi(r_1, r_2, p)}{2\sqrt{L}}\right)\right]^2.
\end{aligned}$$

(26)

The decision variables (22) and (23) are interlinked, and there is no closed form solution; and it is also hard to verify that the solution Q and L satisfy the values of (26) greater than zero. We propose an algorithm to find the optimal solution Q, L, and n.

Algorithm

Step 1 Set $n = 1$

Step 2 Perform steps **2.1** to **2.5**.

Step 2.1 Put $L^{(1)} = 0$ in Eq. (22) and determine the value of $Q^{(2)}$.

Step 2.2 Now substitute the value of $Q^{(2)}$ in Eq. (23) and solve for L, and name it as $L^{(2)}$.

Step 2.3 Utilizing $L^{(2)}$ in Eq. (22) and find the value of $Q^{(3)}$.

Step 2.4 Repeat the steps **2.1** to **2.3** until no change occurs in the values of (Q, L). Denote $(\dot{Q}, \dot{L})$.

Step 2.5 Now substitute $(\dot{Q}, \dot{L})$ into (26). If $\dfrac{\partial^2 \text{ETC}}{\partial Q^2} \times \dfrac{\partial^2 \text{ETC}}{\partial L^2} - \left(\dfrac{\partial^2 \text{ETC}}{\partial Q \partial L}\right)^2 > 0$, then this $(\dot{Q}, \dot{L})$ is the optimal for the given n, denoted by $(Q_{(n)}, L_{(n)})$.

Step 3 Utilizing $(Q_{(n)}, L_{(n)}, n)$ find the $ETC(Q_{(n)}, L_{(n)}, n)$ from (18).

Step 4 Replace n by $n + 1$ and go to **Step 2**.

Step 5 If $ETC(Q_{(n)}, L_{(n)}, n) < ETC(Q_{(n-1)}, L_{(n-1)}, n-1)$, then go to **Step 4**. Otherwise go to **Step 6**.

Step 6 Set the optimal solution of this system as $(Q^*, L^*, n^*) = (Q_{(n-1)}, L_{(n-1)}, n-1)$. And $ETC(Q_{(n-1)}, L_{(n-1)}, n-1)$ is the minimum integrated total expected cost. Stop the algorithm.

5 Numerical Example

In this section, we present a numerical illustration which is almost similar to [9] and [8]. $D = 1000$, $A = 200$, $\pi = 50$, $\pi_0 = 150$, $\sigma = 2$, $\eta_1 = 156$, $\eta_2 = 1$, $\epsilon = 0.7$, $P = 3000$ $C_v = 100,000$, $x = 17,500$, $B = 1500$, $h_b = 5$, $h_v = 4$, $s = 0.5$, $T_r = 25$, $v = 2$, $\theta = 0.3$, $\theta_0 = 0.05$, $\beta = 0.5$; $q = 0.2$, $p = 0.6$, $\beta = 0.6$. In addition, if defective percentage and inspection error follow a uniform distribution with

$$f(y) = \begin{cases} \frac{1}{a}, & 0 \le y \le a \\ 0, & \text{otherwise.} \end{cases}$$

$$f(y') = \begin{cases} \frac{1}{b}, & 0 \le y' \le b \\ 0, & \text{otherwise.} \end{cases}$$

$$f(y'') = \begin{cases} \frac{1}{c}, & 0 \le y'' \le c \\ 0, & \text{otherwise,} \end{cases}$$

then we have

$$\mathbb{E}[y] = \int_0^a y f(y) dy = \int_0^a \frac{y}{a} dy = \frac{a}{2},$$

$$\mathbb{E}[y'] = \int_0^b y' f(y') dy' = \int_0^b \frac{y'}{b} dy' = \frac{b}{2},$$

$$\mathbb{E}[y''] = \int_0^c y'' f(y'') dy'' = \int_0^c \frac{y''}{c} dy'' = \frac{c}{2}.$$

Specifically, if $a = b = c = 0.04$, we have $\mathbb{E}[y] = \mathbb{E}[y'] = \mathbb{E}[y''] = 0.02$. The optimal solutions of example is tabulated in Table 1. The graphical representation of total cost for different value of β and for fixed p is shown in Fig. 1. In Table 2 summary for r_1 and r_2 is tabulated. Sensitivity analysis on ρ and β is tabulated in Table 3.

$$\text{We denote, } D_1 = \frac{\partial^2 \text{ETC}}{\partial L^2} \text{ and } D_2 = \frac{\partial^2 \text{ETC}}{\partial Q^2} \times \frac{\partial^2 \text{ETC}}{\partial L^2} - \left(\frac{\partial^2 \text{ETC}}{\partial Q \partial L} \right)^2.$$

Fig. 1 Graphical representation of the total cost with respect to the number of shipments for different values of β when $p = 0.0$.

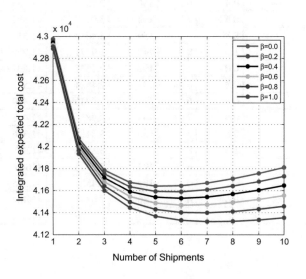

Table 1 Optimal solution for the numerical example

n	Q	L	ETC	D_1	D_2
1	660	2.7768	42,923	12.5562	0.1618
2	414	2.8595	41,981	18.5400	0.5051
3	309	2.8964	41,657	23.9903	1.0904
4	250	2.9178	41,513	29.1298	1.9638
5	211	2.9319	41,446	34.0337	3.1639
6	184	2.9420	41,419	38.7379	4.7234
7	**163**	**2.9495**	**41,416**	43.2735	6.6715
8	147	2.9554	41,427	47.6517	9.0318
9	135	2.9601	41,448	51.8916	11.8264
10	124	2.9640	41,476	56.0003	15.0730

Table 2 Summary for the values of r_1, r_2, and k

p	r_1	r_2	k
0.0	0.14161	0.84161	0.84161
0.2	0.31245	1.01245	0.84013
0.4	0.47102	1.17102	0.84284
0.6	0.61226	1.21226	0.84401
0.8	0.73524	1.43524	0.84282
1.0	0.84161	1.54161	0.84161

Table 3 Sensitivity analysis for different values of β and p

p	β	n	Q	L	ETC	D_1	D_2
	0.0	6	191	2.3291	41,581	65.0638	7.6303
	0.2	6	188	2.5274	41,524	55.2538	6.5688
0.0	0.4	6	186	2.7582	41,463	45.4304	5.4805
	0.6	7	162	3.0445	41,392	39.8306	6.1761
	0.8	7	159	3.4018	41,314	28.9976	4.5846
	1.0	8	140	3.9171	41,220	20.1445	4.0216
	0.0	6	192	2.2625	41,602	68.6635	8.0149
	0.2	6	189	2.4624	41,542	58.3313	6.9052
0.2	0.4	6	186	2.6946	41,479	47.9794	5.7662
	0.6	7	163	2.9824	41,408	42.0559	6.4975
	0.8	7	160	3.3413	41,326	30.6206	4.8271
	1.0	8	140	3.8584	41230	21.2662	4.2358
	0.0	6	192	2.2523	41605	69.2316	8.0757
	0.2	6	189	2.4524	41,545	58.8195	6.9586
0.4	0.4	6	186	2.6847	41,482	48.3891	5.8122
	0.6	7	163	2.9726	41,410	42.4194	6.5502

(continued)

Table 3 (continued)

p	β	n	Q	L	ETC	D_1	D_2
	0.8	7	160	3.3314	41,328	30.8961	4.8685
	1.0	8	140	3.8484	41,231	21.4644	4.2739
	0.0	6	192	2.2262	41,613	70.6806	8.2280
	0.2	6	190	2.4273	41,552	60.0433	7.0902
0.6	0.4	6	187	2.6607	41,488	49.3809	5.9216
	0.8	7	160	3.3101	41,333	31.4852	4.9550
	1.0	8	140	3.8295	41,234	21.8358	4.3435
	0.0	6	191	2.2965	41,591	66.8117	7.8180
	0.2	6	189	2.4955	41,533	56.7500	6.7331
0.8	0.4	6	186	2.7267	41,471	46.6815	5.6214
	0.6	6	186	3.0136	41,400	40.9277	6.3354
	0.8	7	159	3.3713	41,320	29.8079	4.7063
	1.0	8	140	3.8869	41,225	20.7161	4.1312
	0.0	6	191	2.3291	41,581	65.0638	7.6303
	0.2	6	188	2.5274	41,524	55.2538	6.5688
1.0	0.4	6	186	2.7582	41,463	45.4304	5.4805
	0.6	7	162	3.0445	41,392	39.8306	6.1761
	0.8	7	159	3.4018	41,314	28.9976	4.5846
	1.0	8	140	3.9171	41,220	20.1445	4.0216

6 Conclusion

In this article, we have considered the mixture of distribution model in an integrated inventory system. The lead time crashing cost is assumed as a function of L which is negative exponential. An algorithmic procedure is developed to obtain the order quantity, length of the lead time, and the number of shipments. Numerical examples are provided to illustrate the proposed model, and sensitivity analysis is performed for how the different values of β, p affect the entire system. The future work may extend this model for multi-item and fuzzy demand instead of stochastic demand.

Acknowledgements The authors are grateful to the Department of Science and Technology— Science and Engineering Research Board (DST—SERB), Government of India, New Delhi, for providing financial assistance in the form of Fellowship. This research was fully supported by DST-SERB, Government of India, under the grant number DST-SERB/SR/S4/MS: 814/13-Dated 24.04.2014.

References

1. M. Ben-Daya, M. Hariga, Integrated single vendor single buyer model with stochastic demand and variable lead time. Int. J. Prod. Econ. **92**(1), 75–80 (2004)
2. O. Dey, B.C. Giri, Optimal vendor investment for reducing defect rate in a vendorbuyer integrated system with imperfect production process. Int. J. Prod. Econ. **155**, 222–228 (2014)
3. Y.J. Lin, An integrated vendor-buyer inventory model with backorder price discount and effective investment to reduce ordering cost. Comput. Ind. Eng. **56**(4), 1597–1606 (2009)
4. L.I.N. Hsien-Jen, An integrated supply chain inventory model with imperfect-quality items, controllable lead time and distribution-free demand. Yugoslav J. Oper. Res. **23**(1), 87–109 (2013)
5. B.S. Everitt, *Mixture Distributions I*. (Wiley, London, 1985)
6. J.W. Wu, H.Y. Tsai, Mixture inventory model with back orders and lost sales for variable lead time demand with the mixtures of normal distribution. Int. J. Syst. Sci. **32**(2), 259–268 (2001)
7. W.C. Lee, J.W. Wu, W.B. Hou, A note on inventory model involving variable lead time with defective units for mixtures of distribution. Int. J. Prod. Econ. **89**(1), 31–44 (2004)
8. W.C. Lee, J.W. Wu, J.W. Hsu, Computational algorithm for inventory model with a service level constraint, lead time demand with the mixture of distributions and controllable negative exponential backorder rate. Appl. Math. Comput. **175**(2), 1125–1138 (2006)
9. S. Priyan, R. Uthayakumar, Mathematical modeling and computational algorithm to solve multi-echelon multi-constraint inventory problem with errors in quality inspection. J. Math. Model. Algorithms Oper. Res. **14**(1), 67–89 (2015)
10. S. Priyan, R. Uthayakumar, Two-echelon multi-product multi-constraint product returns inventory model with permissible delay in payments and variable lead time. J. Manuf. Syst. **36**, 244–262 (2015)
11. G.K. Bennett, K.E. Case, J.W. Schmidt, The economic effects of inspector error on attribute sampling plans. Naval Res. Logistics (NRL) **21**(3), 431–443 (1974)
12. M. Khan, M.Y. Jaber, A.R. Ahmad, An integrated supply chain model with errors in quality inspection and learning in production. Omega **42**(1), 16–24 (2014)
13. R.J. Tersine, *Principles of Inventory and Materials Management* (North-Holland, New York, 1982)
14. M. Ben-Daya, A. Raouf, Inventory models involving lead time as a decision variable. J. Oper. Res. Soc. **45**(5), 579–582 (1994)
15. D.C. Montgomery, M.S. Bazaraa, A.K. Keswani, Inventory models with a mixture of backorders and lost sales. Naval Res. Logistics Q. **20**(2), 255–263 (1973)
16. L.Y. Ouyang, N.C. Yeh, K.S. Wu, Mixture inventory model with backorders and lost sales for variable lead time. J. Oper. Res. Soc. **47**(6), 829–832 (1996)
17. K. Annadurai, R. Uthayakumar, Reducing lost-sales rate in (T, R, L) inventory model with controllable lead time. Appl. Math. Model. **34**(11), 3465–3477 (2010)
18. S.K. Goyal, An integrated inventory model for a single supplier-single customer problem. Int. J. Prod. Res. **15**(1), 107–111 (1977)
19. P.N. Joglekar, Note comments on "A quantity discount pricing model to increase vendor profits". Manage. Sci. **34**(11), 1391–1398 (1988)

Texture Analysis of Ultrasound Images of Liver Cirrhosis Through New Indexes

Karan Aggarwal, Manjit Singh Bhamrah and Hardeep Singh Ryait

Abstract In health care, one of the most regular diseases is considered that is liver cirrhosis. The mostly accepted method in the identification of liver cirrhosis is by use of ultrasonic images. In this research paper, a method is proposed for identifying the cirrhotic liver through images of ultrasound. The portion of interest has extracted in the cirrhotic and normal ultrasonic images and approved through a radiologist. The cirrhotic liver's recognition is discriminated through new two indexes, i.e., area index and ratio index. The results from the projected new indexes verified its practicability and applicability for recognition of cirrhotic liver.

Keywords Liver cirrhosis · Images of ultrasound · SVM (support vector machine) KNN (K nearest neighbor)

1 Introduction

World Health Organization (WHO) gives the data that millions of citizens worldwide infected by chronic hepatitis C. Due to poor medical awareness, a lot of those will extend liver cirrhosis. The ending phase of chronic hepatopathies is assumed to be cirrhosis which frequently moves toward to hepatocellular carcinoma (HCC) [1]. The diagnosis of the cirrhotic liver disease is best achieved by looking at the rough structure of the liver parenchyma and the aspects of the liver surface such as its unevenness and contours [2, 3]. Ultrasound images are mostly used to differentiate liver cirrhosis.

K. Aggarwal (✉) · M.S. Bhamrah
Electronics & Communication Engineering Department, Punjabi University,
Patiala, India
e-mail: karan.170987@gmail.com

H.S. Ryait
Electronics & Communication Engineering Department, BBSBEC,
Fatehgarh Sahib, India

© Springer Nature Singapore Pte Ltd. 2018
B. Panda et al. (eds.), *Innovations in Computational Intelligence*, Studies in
Computational Intelligence 713, https://doi.org/10.1007/978-981-10-4555-4_7

In the current health system, a radiologist's practice matters to analyze the disease, otherwise it is perceived that additional tool should be extracted in getting better ultrasonic imaging [4]. For that reason, authors have been enthusiastically investigating the quantitative technique to categorize and examine the ultrasonic images of cirrhosis [5, 6]. Biopsy is assumed as best method in discriminating liver diseases, but being in vivo in essence generally not accepted by patients [7].

Ultrasonography is the mainly used method for testing, due to its noninvasive and nonradioactive nature [9]. Here, it has been recognized by the radiologist that the body tissue texture gives an important visual parameters in categorization of outcome in their radiological considerable [10]. Texture of image can be described such as diminutiveness, randomness, stiffness, evenness, surface granulation, or hummocky [11]. In the ultrasonic images of liver, the spatial and intensity collection of pixels defines all these features [12, 13]. This irregularity is due to eruptions/scares on the liver surface [14–16]. The normal liver has regular surface, and cirrhotic has irregular surface as shown in Fig. 1a, b.

The mostly seen pattern of liver disease is micronodular cirrhosis [17–20]. Moreover, normal liver structural design spoiled by scar matter that makes a group of linked tissue combining the perivenous and periportal areas [21–24]. This is generally accepted that eventually micronodular cirrhosis will transformed into cirrhosis of macronodular type. The succession may move toward whole cirrhosis [25–28]. This becomes a difficult work in recognizing the transform when ultrasonic inspection happens. Radiologist desires a serious examination to distinguish between these two [29–31]. If an analytic tool be able to develop based on the theory of image processing, disease recognition is achievable early.

The proposed effort is an attempt to identify and categorize cirrhotic liver based on pattern retrieved through ultrasonic images. In the past time, GLCM, SGLDM, and Law mask has increased attention in image and computer vision [32–35]. That is why some new indexes are required to differentiate liver cirrhosis from normal liver. So, two indexes are identified through experimentation on the ultrasound images of both liver cirrhosis and normal liver. To validate these indexes, support vector machine is used to classify these indexes.

Fig. 1 a Normal liver, **(a)**   **(b)**
b cirrhotic liver:
by NIDDKD [8]

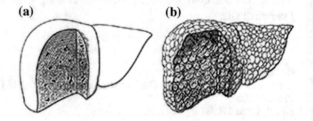

2 Methodology

The ultrasonic images were taken through radiology department in care of a skilled radiologist. The ultrasonic images were of 381 by 331 sizes each and all were in .jpg format. These images were of two types, i.e., one was normal and other was cirrhotic liver. One ultrasound machine was used to take all the images because change in dimensions and texture values affect if this acquired via separate machines. The specification of ultrasonic probe was 15 cm depth and 5 MHz frequency. Radiologist also depicted the area in the ultrasonic images that related to cirrhosis as shown in Fig. 2 for estimation. Region of interest (ROI) has chosen as templates of 50 × 50 sizes through both normal and cirrhotic ultrasonic images as in Fig. 2a–d. For that reason, total of 20 ROIs from 20 normal liver images and 44 ROIs from 44 cirrhosis liver images are taken. Then, the proposed approach was implemented on it.

Methodology adopted for the detection depends upon area cover by maximum occurrence of the intensities in the image. Here area of white intensity was computed which conveys the information about the identical pixels existing in image.

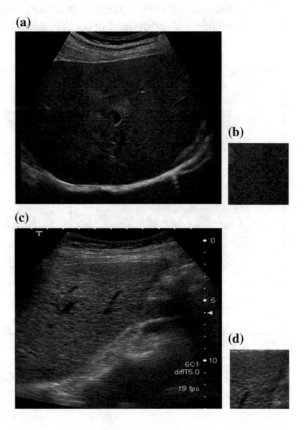

Fig. 2 **a** Normal liver ultrasound image, **b** region of interest in normal liver, **c** cirrhosis liver ultrasound image, **d** region of interest described by radiologist in cirrhosis liver

Understanding taxonomy is important to predict the values of these new indexes. If there is a regular texture in an image as shown in Fig. 3a, then regularity must be in its gray-level distribution. Hence, all the intensities in the image are occurring equally, so when replacement of the maximum occurrence of the intensities by 255 happens, then the almost complete image becomes white, i.e., 255 as shown in Fig. 3c and area of white in the image becomes greater and ratio becomes smaller. Similarly, if there is an irregular texture in an image as in Fig. 3b, then nonuniform allocation of its grayscale must be there; therefore, the intensities in the image are occurring nonuniformly, so when replacement of the maximum occurrence of the intensities by 255 happens, then the intensities that occur maximum in the image become white, i.e., 255, and rest of the intensities becomes black as shown in Fig. 3d and area of white in the image becomes smaller and ratio becomes greater.

This very property makes the decisive factor for detection of regular pattern or irregular pattern. Figure 1 clearly depicts the change in the taxonomy of liver surface due to cirrhosis which can be easily detected by the proposed decisive factor. The flow chart described the complete methodology as shown in Fig. 4.

In the present study, this ideology has been applied for differentiating the cirrhotic liver through ultrasound images. As shown in flow chart, 30 intensities that occur maximum were separate out and replace all these 30 intensities by intensity 255, i.e., white and rest of the intensities replace by intensity 0, i.e., black. Many hit and trial methods were performed to set this value, i.e., 30 intensities. Then, new

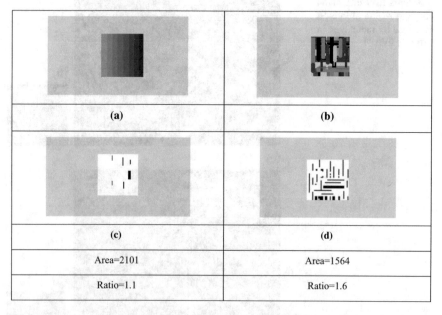

Fig. 3 **a** Regular texture image, **b** irregular texture image, **c** result after proposed work on regular texture image, **d** result after proposed work on irregular texture image

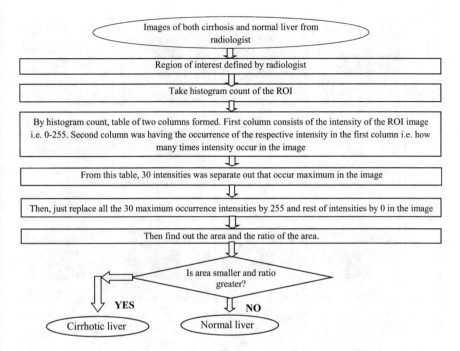

Fig. 4 Flow chart of proposed method

indexes (area and ratio) were computed that used to differentiate between a regular and an irregular pattern. To validate these indexes, support vector machine is used to classify these indexes.

3 Result

Region of interest (ROI) of size 50 × 50 has been taken as template from each ultrasound image. Accordingly, total of 20 ROIs from 20 normal liver images and 44 ROIs from 44 cirrhosis liver images are taken.

When this new approach is applied on the normal liver templates, i.e., regular pattern as 4 normal templates shown in Fig. 5c(i–iv), all the intensities are uniform. Taking 30 intensities that is occurring maximum in the image and replaces these all intensities by intensity 255 or white and rest of the intensities replace by intensity 0, i.e., black, so new image formed as shown in Fig. 5d(i–iv). Many hit and trial methods are performed to set this value, i.e., 30 intensities. Similarly, this approach is applied on the cirrhosis liver templates, i.e., irregular pattern as 4 cirrhosis templates shown in Fig. 5a(i–iv), all the intensities are nonuniform. Taking 30 intensities that is occurring maximum in the image replaces these all intensities by

	Cirrhosis 1	Cirrhosis 2	Cirrhosis 3	Cirrhosis 4
	cirrhosis 1	cirrhosis 2	cirrhosis 3	cirrhosis 4
	u(i)	u(ii)	u(iii)	u(iv)
	v(i)	v(ii)	v(iii)	v(iv)
Area	1320	1.0921e+03	1.2754e+03	1.5504e+03
Ratio	1.8999	2.289	1.96	1.6124
	Normal 1	Normal 2	Normal 3	Normal 4
	normal 1	normal 2	normal 3	normal 4
	w(i)	w(ii)	w(iii)	w(iv)
	x(i)	x(ii)	x(iii)	x(iv)
Area	2.2038e+03	2.1626e+03	2.2973e+03	2.3728e+03
Ratio	1.134	1.156	1.088	1.053

Fig. 5 $u(i–iv)$ templates for cirrhosis liver and $v(i–iv)$ result images after applied proposed method on cirrhosis liver templates and $w(i–iv)$ templates for normal liver and $x(i–vi)$ result images after applied proposed method on normal liver templates

intensity 255 or white and rest of the intensities replace by intensity 0, i.e., black, so new image formed as shown in Fig. 5b(i–iv).

After getting these new resultant images, the area of the white intensity was computed, and then ratio of area of white intensity and complete ROI image was computed.

$$\text{Ratio} = \frac{\text{area of complete ROI image}}{\text{area of white intensity occupied in the resultant image}}$$

In normal liver, due to regular pattern, almost all the intensities cover under 30 maximum occurrence intensities, so they all become white and very few become black, white area becomes larger than black, and ratio becomes small. On the other side, in cirrhotic liver, due to irregular pattern, some intensities cover under 30 maximum occurrence intensities, so they become white and rest becomes black.

In this, black area is large as compared to white, so white area is small as compared to normal liver and ratio becomes large as compared to normal liver.

For statistical analysis, most acknowledged classifiers, i.e., support vector machine (SVM) and K nearest neighbor (KNN) are used. SVM are fewer prone to over fitting, and KNN does not need any assumption about the data allocation.

In support vector machine (SVM) classifier, the normal liver is depicted by pink '+' symbol and the cirrhotic liver is depicted by blue '+' symbol. The line divides an area of the cirrhotic and normal liver, and the circle on the plus symbol shows the support vectors. In this, only two out of 44 blue '+' falls in the normal region and only one out of 20 pink '+' falls in cirrhotic region as in Fig. 6a. So accuracy is 95.3%, sensitivity is 95.45%, and specificity is 95% as shown in Table 1.

In KNN classifier graph, normal liver images are depicted by blue dots, and cirrhotic liver is depicted by red dots. In this, accumulation of blue dots is one side and red dots are other side. The black cross sign is the differentiation boundary between cirrhosis and normal regions. There are only two blue dots out of 20 falls in the red regions and two red dots out of 44 falls in the blue regions as in Fig. 6b. So accuracy is 93.8%, sensitivity is 95.4%, and specificity is 90% as shown in Table 1.

Table 1 Classifiers performance	Support vector machine	K nearest neighbor
Accuracy in %	95.3	93.8
Sensitivity in %	95.45	95.4
Specificity in %	95	90

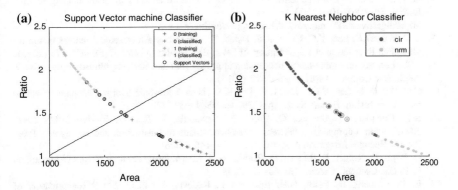

Fig. 6 **a** Graph of SVM classifier, **b** graph of KNN classifier

4 Conclusions

In this study, new indexes have been identified for identification of surface in cirrhotic and normal liver ultrasound images. The proposed method was checked on a data of 44 cirrhosis templates and 20 normal templates. The best outcome is achieved in identifying cirrhotic liver. Support vector machine classifier has provided the 95.3% accuracy, sensitivity is 95.45%, and specificity is 95%. Similarly, K nearest neighbor (KNN) has given the accuracy of 93.8%, sensitivity is 95.4%, and specificity is 90%. In the future work, the different stages of cirrhotic liver can be classified by using this technique.

References

1. J. Virmani, V. Kumar, N. Kalra, N. Khandelwal, Prediction of cirrhosis based on singular value decomposition of gray level co-occurence matrix and a neural network classifier, in *Proceeding of IEEE Conference: Developments in E-systems Engineering*, pp. 146–151, Dec 2011
2. K. Fujino, Y. Mitani, Y. Fujita, Y. Hamamoto, I. Sakaida, Liver cirrhosis classification on m-mode ultrasound images by higher-order local auto-correlation features. J. Med. Bio. 3(1), 29–32 (2014)
3. K. Aggarwal, M.J. Bhamrah, H.S. Ryait, The identification of liver cirrhosis with modified LBP grayscaling and Otsu binarization. Springerplus 5(1), 1–15 (2016)
4. J.W. Jeong, S. Lee, J.W. Lee, D.S. Yoo, S. Kim, The echotextural characteristics for the diagnosis of the liver cirrhosis using the sonographic images, in *Proceeding of IEEE Conference: Engineering in Medicine and Biology Society*, pp. 1343–1345, 2007
5. D. Mitrea, S. Nedevschi, R. Badea, The role of the multiresolution textural features in improving the characterization and recognition of the liver tumors, based on ultrasound images, in *International Symposium on Symbolic and Numeric Algorithms for Scientific Computing*, pp. 192–199, Sept 2012
6. R.M. Hawlick, Statistical and structural approaches to texture. Proc. IEEE Conf. 67(5), 786–808 (1979)
7. G. Castellano, L. Bonilha, L.M. Li, F. Cendes, Texture analysis of medical images. Clin. Radiol. 59, 1061–1069 (2004)
8. http://digestive.niddk.nih.gov/ddiseases/pubs/cirrhosis_ez
9. C.C. Wu, W.L. Lee, Y.C. Chen, K.S. Hsieh, Evolution-based hierarchical feature fusion for ultrasonic liver tissue characterization. IEEE J. Bio. Health Inform. 17(5), 967–976 (2013)
10. W.L. Lee, An ensemble-based data fusion approach for characterizing ultrasonic liver tissue. Appl. Soft Comput. 13(8), 3683–3692 (2013)
11. C.C. Wu, W.L. Lee, Y.C. Chen, C.H. Lai, K.S. Hsieh, Ultrasonic liver tissue characterization by feature fusion. Expert Syst. Appl. 39(10), 9389–9397 (2012)
12. K.L. Caballero, J. Barajas, O. Pujol, N. Salvatella, P. Radeva, In-vivo IVUS tissue classification a comparison between normalized image reconstruction and RF signals. Prog. Pattern Recogn. Image Anal. Appl. 4225, 137–146 (2006)
13. O. Pujol, D. Rotger, P. Radeva, O. Rodriguez, J. Mauri, Near real-time plaque segmentation of IVUS. Comput. Cardiol. 30, 69–72 (2003)
14. E. Brunenberg, O. Pujol, B.H. Romeny, P. Radeva, Automatic IVUS segmentation of atherosclerotic plaque with stop & go snake. Med. Image Comput. Comput. Assist. Interv. 4191, 9–16 (2006)

15. T. Ojala, M. Pietikäinen, D. Harwood, A comparative study of texture measures with classification based on featured distribution. Pattern Recogn. **29**(1), 51–59 (1996)
16. A. Hadid, M. Pietikainen, T. Ahonen, A discriminative feature space for detecting and recognizing faces, in *Proceeding of International Conference: Computer Vision Pattern Recognition*, pp. 797–804, 2004
17. D. Grangier, S. Bengio, A discriminative kernel-based approach to rank images from text queries. IEEE Trans. Pattern Anal. Mach. Intell. **30**(8), 1371–1384 (2008)
18. W. Ali, F. Georgsson, T. Hellstrom, Visual tree detection for autonomous navigation in forest environment, in *Proceeding of IEEE Conference: Intelligent Vehicles Symposium*, pp. 560–565, June 2008
19. L. Nanni, A. Lumini, Ensemble of multiple pedestrian representations. IEEE Trans. Intell. Transp. Syst. **9**(2), 365–369 (2008)
20. T. Maenpaa, J. Viertola, M. Pietikainen, Optimising colour and texture features for real-time visual inspection. Pattern Anal. Appl. **6**(3), 169–175 (2003)
21. M. Turtinen, M. Pietikainen, O. Silven, Visual characterization of paper using Isomap and local binary patterns. IEICE Trans. Inf. Syst. **E89D**(7), 2076–2283 (2006)
22. M. Heikkila, M. Pietikainen, A texture-based method for modeling the background and detecting moving objects. IEEE Trans. Pattern Anal. Mach. Intell. **28**(4), 657–662 (2006)
23. V. Kellokumpu, G. Zhao, M. Pietikainen, Human activity recognition using a dynamic texture based method, in *Proceeding of British Machine Vision Conference* (Leeds, UK, 2008)
24. A. Oliver, X. Llado, J. Freixenet, J. Marti, False positive reduction in mammographic mass detection using local binary patterns, in *Proceeding of Medical Image Computing and Computer-Assisted Intervention Conference*, pp. 286–293, 2007
25. S. Kluckner, G. Pacher, H. Grabner, H.A. Bischof, 3D teacher for car detection in aerial images, in *Proceeding of IEEE International Conference: Computer Vision*, pp. 1–8, 2007
26. A. Lucieer, A. Stein, P. Fisher, Multivariate texture-based segmentation of remotely sensed imagery for extraction of objects and their uncertainty. Int. J. Remote Sens. **26**(14), 2917–2936 (2005)
27. R. Rodríguez, A strategy for blood vessels segmentation based on the threshold which combines statistical and scale space filter: application to the study of angiogenesis. Comput. Methods Programs Biomed. **82**(1), 1–9 (2006)
28. D.Y. Huang, C.H. Wang, Optimal multi-level thresholding using a two-stage Otsu optimization approach. Pattern Recogn. Lett. **30**(3), 275–284 (2009)
29. A. Tamim, K. Minaoui, K. Daoudi, H. Yahia, A. Atillah, D. Aboutajdine, An efficient tool for automatic delimitation of moroccan coastal upwelling using SST images. IEEE Geosci. Remote Sens. Lett. **12**(4), 875–879 (2015)
30. P. Filipczuk, T. Fevens, A. Krzyżak, Computer-aided breast cancer diagnosis based on the analysis of cytological images of fine needle biopsies. IEEE Trans. Med. Imaging **32**(12), 2169–2178 (2013)
31. T. Ojala, M. Pietikainen, T. Maenpaa, Multiresolution gray-scale and rotation invariant texture classification with local binary patterns. IEEE Trans. Pattern Anal. Mach. Intell. **24**(7), 971–987 (2002)
32. N. Otsu, A threshold selection method from gray-level histograms. IEEE Trans. Syst. Man Cybern. **9**(1), 62–66 (1979)
33. A.K. Chaou, A. Mekhaldi, M. Teguar, Elaboration of novel image processing algorithm for arcing discharges recognition on HV polluted insulator model. IEEE Trans. Dielectr. Electr. Insul. **22**(2), 990–999 (2015)
34. S. Murala, Q.M. Jonathan, Local mesh patterns versus local binary patterns: biomedical image indexing and retrieval. IEEE J. Bio. Health Inform. **18**(3), 929–938 (2014)
35. Y. Hu, C. Zhao, A local binary pattern based methods for pavement crack detection. J. Pattern Recogn. Res. 140–147 (2010)

Neoteric RA Approach for Optimization in ZRP

Mehta Deepa, Kashyap Indu and Zafar Sherin

Abstract To address the challenges brought by digital deluge, the progressive and self-organized communication system is required to make a better connected world in which anything can be shared anywhere. MANET protocols, like ZRP, provide such communication system by providing a mobile wireless network with nodes moving freely in any direction/on any track with the option of altering its connection. This paper focuses on the neoteric performance optimization approach for ZRP by pointing out the gaps of study in existing routing mechanism of ZRP. This gives rise to the neoteric strategy of route aggregation (RA) also called route summarization or supernetting which is an approach that supplants a group of routes by a lone more common route, utilized for optimized performance of routing methodology of ZRP, hence will further improve QOS of ZRP.

Keywords MANET · QOS · ZRP · RA neoteric strategy

1 Introduction

Mobile ad hoc networks [1] justifying their name are wireless, notably designated systems comprising of randomly moving nodes, as a result network topology changes frequently and unpredictably. Thus, mobility of nodes results in a need of developing different efficient routing protocols. Ad hoc networks are turning out to be increasingly popular and are garnering immense interest owing to easy,

M. Deepa (✉) · K. Indu
Department of Computer Science & Engineering (FET), Manav Rachna International
University, Faridabad, India
e-mail: deepa.mehta12@gmail.com

K. Indu
e-mail: indu.fet@mriu.edu.in

Z. Sherin
CSE, SEST, Jamia Hamdard University NCR, New Delhi, India
e-mail: Sherin_zafar84@yahoo.com

© Springer Nature Singapore Pte Ltd. 2018
B. Panda et al. (eds.), *Innovations in Computational Intelligence*, Studies in
Computational Intelligence 713, https://doi.org/10.1007/978-981-10-4555-4_8

infrastructure-less communication. With this, the ad hoc networks also have to withstand various obstacles in the due course [2]. The highly mobile limited-powered asymmetrically connected nodes keep varying the topology of the network, thus devising a coherent routing protocol becomes absolutely essential. Various past researches have led to the following categorically divided protocols:

1. Proactive
2. Reactive
3. Hybrid

Proactive routing protocols [3] strive to maintain current distance estimates with other nodes of the network utilizing a large section of bandwidth. Due to the rapidly varying topology the route may have to be refreshed often thus squandering the meager bandwidth. Reactive routing protocols [4] explicitly locate packet forwarding routes on exigency thus saving bandwidth, but the route discovery process introduces delays and also stems immoderate traffic problems. Hybrid protocols [5] integrate the advantages of both the above organically and try to obtain a higher performance. The hybrid protocols are capable of rendering higher scalability than its counterparts. Ingenuity of hybrid protocols is that they make an effort to get rid of the single point failures in the network and bottleneck problems by permitting all the nodes to route and forward data in case of unavailability of preferred path.

In an attempt to keep an eminent topological information, each node records the path existing from one node to another node in every zone. This leads to redundant routes out of which only the efficient one is used to forward the data, but the redundant and practically volatile information stored in the routing table requires more memory and also amount of power consumption increases. MANET suffers from continually changing topology, scarce transmission power, and disproportioned links owing to asymmetry; this originates a need for optimization in hybrid routing protocols. To provide the performance optimization by overcoming the above limitations, development of various neoteric strategies becomes important. In this paper, we aim to focus on such a strategy called route aggregation (RA) [6] which help in eliminating the routing overhead at each node. The upcoming sections of the paper will focus ZRP routing strategy and the application of neoteric strategy of RA for QOS optimization, discussion, and conclusion.

2   Related Works

2.1   ZRP

A dispensed grid of independent high mobility nodes with preparedness to interact amongst one another wirelessly forms an ad hoc network. The participating nodes are highly ambulant in nature thus leading to high speed changes in the topology of the network which in turn coerces the need for each router to discover the most appropriate route to the concerned destination on its own as the mobility makes the utilization of common routing tables perpetuated at the routers infeasible for use. In a MANET, with each node acting as a router forwarding data, most of the communication happens within the closer nodes. In case of any variation in the topology due to any node joining or exiting a network, the primarily affected nodes are the neighbor nodes within the locality and far away nodes have limited influence. ZRP [7] efficiently utilizes the same concept by keeping a comprehensive record of neighboring nodes by dividing the network into zones.

In ZRP, whenever a source node initiates a packet to be sent, it is started by checking its own local zone for the destination by using IARP [8] and the packet is routed through the existing proactively maintained route. But in case the source is unable to find the destination in its own zone, the packet is transferred reactively. At this stage, route is requested by sending a route request packet to adjoining node located at its zone periphery with support from BRP [9]. Node receiving the request if aware of the destination node answers using a route reply packet back to the sender, else the continuity of border-casting propagates the route request throughout the network. With the movement of nodes, the links will break and new ones will be formed thus making maintenance of route a prime requirement.

ZRP is indistinctive in nature, each node's home zone is separated from the overall topology of whole of the network, and thus, it works on the application of distinct approaches. Every node singly forms its own locale which becomes its own zone for routing. The formation of zone happens when a group of nodes distant from the node in context is not more than the set value called the radius of the zone. The nodes with distance from the node in context is equal to the radius of the zone are called peripheral nodes. These locales are called zones (justifying their name); every node might be a member of numerous overlapping zones, and different zones may have different sizes. The radius of the zone is calculated in terms of number of hops from the concerned node. The concept of splitting the network into imbricated, variably sized zones, ZRP eludes the formation of a hierarchy into the network thus reducing the overhead that is increased in case of maintenance of this map. Instead, the network is flat, with a possibility of route optimization with the detection of overlapping zones.

The nodes falling on the zone perimeter (i.e., with number of hops to the node = r) are denoted as peripheral nodes, nodes with number of hops to the node < r are interior nodes. In an attempt to construct a routing zone, node requires information about its neighbors and to ascertain its peripheral nodes. Neighbor discovery protocol (NDP) [10] helps the node in gathering information about its immediate neighbors. Neighbor Discovery Protocol counts on the circulation of "hello" messages by each and every node. On receiving response to the hello message, the node records the straight point-to-point link with that particular neighbor. The NDP freely chooses nodes on different basis, which may be greater signal strength or the frequency/delay of beacons etc. After gathering neighboring information about the route, node still retains current information about the neighbors with the help of discovery messages which are broadcasted at regular intervals. Due to the dynamic locale largely affecting node's routing behavior more in contrast to any change on any other side of the network, proactive, table-driven protocol is used known as IARP [8]. The node constantly requires an updated information about the peripheral nodes and about the nodes at one-hop distance. IARP empowers the removal of the redundant routes, choosing the route with least hops and also avoiding faulty links. Thus, achieving optimized local route.

IARP must be able to render support for unidirectional links amongst the local nodes. Proactive strategy application makes the discovery of local routes very efficient and readily available. IARP is restricted to only the locale hence limited use of scarce bandwidth.

- Every node retains a table to maintain routes for its routing zone, making itself capable of finding route to any destination node within its own zone provided by the table entries.
- Each node at regular intervals of time keeps broadcasting messages similar to hello beacon for the zone notification.
- After one hop, any hello message ceases to exist as it makes it to the neighbor by then.
- A message for zone notification fails to exist after it reaches the neighbor of the node at x hops.

On the receipt of this message, each node subtracts the message hop count by 1 during transmission to its neighbors. ZRP adroitly amalgamates the merits of reactive along with proactive strategies. An ad hoc network each node operates by steering the traffic to the next hop router. Therefore, ZRP incorporates proactive approach within a zone with each node as the center of its own zone. Thus, reducing the amount of proactive data to be maintained. For the nodes outside the zone the strategy is reactive. Thus reducing the number of nodes to be queried for ascertaining the route. Figure 1 depicts ZRP Architecture [11] where the routing zone formation happens as follows: Neighbor Discovery Protocol (NDP): broadcast "hello" beacons intermittently. After receiving a "hello" the node updates its neighbor table. The tables are refreshed by eliminating the entries of neighbor with no beacon within determined time.

The routing zone formation happens as follows: Neighbor Discovery Protocol (NDP): broadcast "hello" beacons intermittently. On receiving a "hello," the node restores its neighbor table. The tables are refreshed by eliminating the entries of neighbor with no beacon within determined time. NDP functionality is either rendered by MAC layer or proffered by IARP as shown in Fig. 2.

Classified nodes within the zone of ZRP are:

(i) Interior nodes: the nodes with minimum distance to the central node less than the radius of the zone.
(ii) Peripheral nodes: the nodes whose distance to the central node is equal to the radius of the zone.

Figure 2 shows the routing zone of X1. The nodes within the routing zone are X2 to X12 but not Y because they are distant from X1 by either 2 hops or less,

Fig. 1 Architecture of ZRP

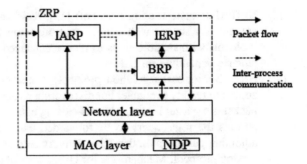

Fig. 2 Example of routing zone with radius = 2

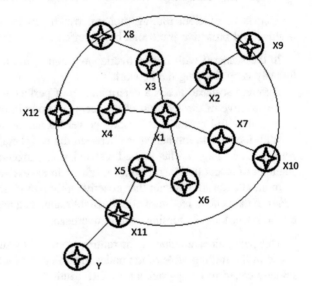

whereas Y is 3 hops away so is out of the zone. But nodes X2 to X6 are called interior nodes, whereas nodes X8 to X12 are called peripheral nodes.

2.2 Constituents of ZRP

ZRP includes two routing constituents, IARP which is the regional proactive constituent and IERP is the comprehensive reactive constituent, which are further discussed below:

(i) IARP intrazone routing protocol: This regional proactive routing constituent is not a well-defined protocol instead is explicated from proactive routing protocols with modification limiting its range of route updating to the limited routing zone of the node.

(ii) IERP (Interzone routing protocol) [12]: This comprehensive reactive constituent is also not well defined and are explicated from reactive routing protocols with modifications to render enhanced route discovery and maintenance depending on IARP.

(iii) BRP (Border-cast resolution protocol): Since in ZRP the route within the zones are readily available. For discovering a route globally broadcasting, the packet is a waste. Hence, a border-casting concept is utilized. Border-casting permits the route query to be forwarded using multicasting to a group of adjoining peripheral nodes. This delivery service is provided by border-cast routing protocol. Mechanisms for Query control support ZRP in performing querying efficiently even though the zones overlap heavily.

ZRP is an efficient routing protocol which proves to be better than the pure reactive and proactive protocols, but it suffers from a few limitations [13]:

In the scenario with highly overlapping zones, the route request packet superfluously keep flooding the network.

Different scenario needs a determination of perfect zone radius.

Irrespective of the location of the recipient node in the zone the forwarding of packets uses full power thus wasting power for nearby nodes.

With increase in the interspace between the peripheral and the source node, the zone gets extended. The limited transmission range of sender node provokes attempts to detect peripheral nodes resulting in excessive use of bandwidth.

In an attempt to maintain the proactive information of entire set of neighboring nodes in the zone, a record of all redundant routes to a particular destination is kept by the node, thus, increasing routing overhead.

This paper aims at reducing the routing overhead by suppressing the information stored in the routing table of the nodes regarding the routes by applying a neoteric strategy called route aggregation to ZRP routing.

3 Neoteric RA Approach for Optimization in ZRP

With rising interest toward MANET owing to the ease of infrastructureless deployment of network to facilitate communication, there arises a need for large scale networks. Increasing scalability [14] increases the memory requirement at the node in order to accommodate the additional routing information, and thus, the capability of system to subsist and operate under augmented load becomes an important requirement. To incorporate more scalability in a network, the routers must be capable of recapitulating comprehensive routing information about closer nodes and crude information about farther nodes in their routing tables. In a chase to retain the information regarding routing about all the neighboring nodes, the size of the routing table increases and so does the memory requirement whilst not all the stored information is necessary; thus, there needs to be a mechanism to eliminate the redundant routes, thus suppressing the table size. One such practical approach is called route aggregation [15] which aims to eliminate dispensable entries of the routes within the network from the routing table. Route aggregation (also called route summarization or super netting) is an approach that supplants a group of routes by a lone route.

The advantages of applying RA [16] in ZRP will help in curbing the routing table size, reducing the number of route advertisements sent by the nodes, allot prevailing set of addresses sensibly, formulate an efficient routing process, reduce the burden on memory resources, and isolate the changes in topology. Figure 3 depicts the route aggregation methodology. With the increase in the amount of networks attached to the internet, its routing system started to face significant scalability issues. The Border Gateway Protocol (BGP) [17] being absolutely main protocol responsible for exchange of information amongst the autonomous systems, and hence, it is found that its routing table is growing at a very fast rate. To contain this, problem route aggregation (RA) plays a very crucial role.

RA [18] promulgates a lone common route in spite of announcing all the child routes coming within a general parent prefix, thus helping in suppressing the BGP routing table size as illustrated in Fig. 3. Each router R_i $(0 \leq i \leq 15)$ is connected to directly with an interface having the IP prefix as 172.1.i.0/24. So the router Q has

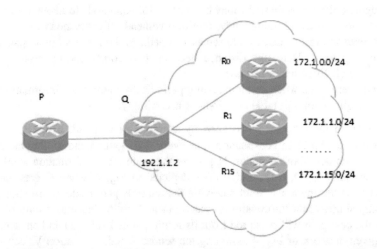

P's routing table		P's routing table	
Destination	Next Hop	Destination	Next Hop
172.1.0.0/24	192.1.1.2	172.1.0.0/20	192.1.1.2
172.1.1.0/24	192.1.1.2		
. . . . .	. . . . .		
172.1.15.0/24	192.1.1.2		

Fig. 3 Route aggregation rendering router Q to combine multiple routes into single route **a** without aggregation at router Q, **b** with route aggregation at router **Q**

16 entries in its routing table 172.1.0.0/24, 172.1.1.0/24, ..., 172.1.15.0/24. In the absence of RA, the routing table will look like as in Fig. 3a. But RA allows to aggregate the routes to be replaced by a single prefix for destination and announce only 172.1.0.0/20 to the router P as shown in Fig. 3b.

4 Illustrative Examples

Below describes the scenario for neoteric RA-based optimization approach. In an area of 1000×1000, we take 100 nodes, randomly at some locations:

- First task is that of identification of nodes in the entire network.
- With the help of periodic HELLO packets, the neighbor nodes are able to communicate to each node, individual creation of one's own neighborhood or zone.
- Each node forms a zone for routing with radius 2 (hops).
- Division of nodes within a zone as peripheral and interior nodes. Nodes lying on the periphery of the zone are called peripheral nodes. The nodes inside the zone periphery are called interior nodes.
- Route aggregation: Several paths may be existed from one node to another node in every zone. To avoid redundancy, routing overhead on every node in zone, the path with shortest distance is chosen as best path. And remove all other paths from routing table. Here, the distance between two corresponding nodes is calculated with Euclidian's formula.
- Each and every node when communicating periodically gathers the information of their neighbor routing table and updates accordingly.

Every node on the network is assigned by a unique network id. In the given below network, sender H and destination J, as every node gathers the routing table information of its neighboring nodes, H gathers the route table information available at C, similarly C, D, S, F, and E does depicted in Fig. 4, when C gets the multiple paths to destination J. After receiving the multiple path route information, C stores only single route information to reach out to J, following that if multiple paths exit, the best path will be stored from its routing table (Table 1) and chosen.

In the given network of Fig. 4 assuming the sender S and destination V, common ZRP routing and neoteric RA-based routing are illustrated below:

Common ZRP Routing

1. Firstly, S searches with the interior nodes A, B, C, D, E, and F. If destination V is found, packet will be delivered, else request forwarded to peripheral nodes.
2. Now, peripheral nodes G, H, I, J, and P are searched for the destination, if found delivers the packet, if not found interzone routing is done.
3. Now, a request to find the destination V is forwarded to another zone (I zone) through the peripheral nodes which connected to the other zone.

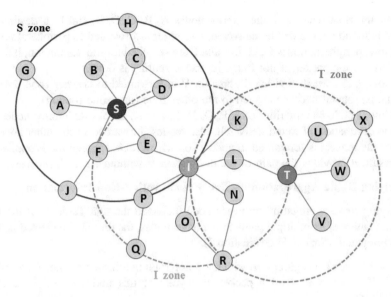

Fig. 4 A sample ZRP scenario

Table 1 Applying route aggregation at one node, i.e., T

Source	Destination	Description	Forwarding path	Aggregated path
S	V	Checks all its interior nodes for V, if found delivered, else forwarded to another zone through I s–c–h, s–d, s–a, s–b, s–g, s–f, s–j, s–e	S–E–I	
		I node checks all its interior nodes for V, if found delivered, else forwarded to another zone	I–L–T	
		On receiving the packet at T node, its directed to V node directly, as the path is aggregated at T node		T $\sum$ (U, X), T $\sum$ (W), and T $\sum$ (V),

4. Step 1, 2, and 3 are repeated until the destination is found. The routing within the zone is a table-driven routing, out of specific zone is on-demand.

Applying Route Aggregation at One Node, i.e., T

Applying route aggregation on node T (depicted in Table 1) since there is only 1 route from L to T. So routing table of T consist all the routing information of U, X, W, and V nodes.

1. Firstly, **S** searches with the interior nodes A, B, C, D, E, and F. If destination **V** is found packet will be delivered, else request forwarded to peripheral nodes.
2. Now, peripheral nodes G, H, I, J, and P are searched for the destination, if found delivers the packet, if not found interzone routing is done.
3. Now, a request to find the destination V is forwarded to another zone through the peripheral node connected to the other zone (interzone routing).
4. Now, I node has interior nodes P, O, N, L, and K, checks its interior nodes for the destination, if found delivered, else request forwarded to the other zone.
5. So the request is forwarded to node T, on which route aggregation is applied as result; it provides the path of the destination V without a search process.

Applying Route Aggregation on Every Node with 2-Hop Information

Applying route aggregation on every node (depicted through Table 2) of the network, if there exist multiple paths to the other nodes, the directly connected path is an aggregated. (Sender H destination J).

1. As the node H is aggregated with the 2-hop routing information, i.e., C, S and C, D, if it wants to send a packet to anyone of this nodes packet is directly delivered.
2. In case H sends a packet to J, firstly it looks up its routing table if there exist a path it is directly delivered; otherwise, it's forwarded to the next node.
3. The next node does the same as the H node, i.e., the C and forwards the packet to S node not to D; as in its aggregation table of 2 hops, S is directly connected to C.
4. When packet is received by node S, its checks its aggregated routes in its routing table and direct the packet to J, as S node knows the path to the J.
5. If J is not found in S aggregated path, packet is forwarded to next nodes.
6. The next node does the same as the previous nodes, and this process is carried on until the packet is delivered.

The final aggregated routing table (Table 3) thus shows the routes are aggregated and are supplanted by a single the best route thus saving the memory resources.

Table 2 Aggregated routing table

Source	Destination	One-hop neighboring nodes	Two-hop neighboring nodes	Aggregated path
H	J	H–C, C–D, C–S, D–S	H–C–S, H–C–D	$H \sum (C, S)$ and $H \sum (C, D)$, H–C–D, H–C–S
		C–S, C–D	C–S–F, C–S–E, C–D–S	$C \sum (S, F)$ and $C \sum (S, E)$, C–S–F, C–S–E
		S–F, S–E	S–F–J, S–F–I	$S \sum (F, J)$, and $T \sum (F, I)$. S–F–J, S–F–I

Table 3 Aggregated routing table

Source	Destination	Paths available	Aggregated path
H	J	H–C–S–F–J H–C–D–S–F–J H–C–D–S–E–F–J H–C–S–E–F–J	H–C–S–F–J
G	H	G–A–S–C–H G–A–S–D–C–H G–A–S–F–E–S–C–H G–A–S–F–E–S–D–C–H G–A–S–E–F–S–C–H G–A–S–E–F–S–D–C–H	G–A–S–C–H
H	B	H–C–S–B H–C–D–S–B H–C–S–E–F–S–B H–C–S–F–E–S–B H–C–D–S–E–F–S–B H–C–D–S–F–E–S–B	H–C–S–B

5 Conclusion and Future Scope

In this paper, a neoteric approach has been utilized to overcome the various limitations of routing in ZRP. The proposed research study uses route aggregation as a neoteric and effective optimization strategy for finding the best possible route simultaneously enhancing the QOS parameter for ZRP. Route aggregation follows a summarization methodology which allows changes in the routing methodology of ZRP to enhance and optimize its performance which is illustrated well through above discussion and routing examples. Therefore, RA has been applied in BGP, OSPF, and IPV6, proposing this strategy for MANETs is a neoteric contribution in this area of research which will be further contributed by simulation results.

References

1. Mobile Ad-Hoc Networks (MANET) charter, IETF. http://www.ietf.org/html.charters/manet-charter.html
2. S. Corson, J. Macker, Mobile ad hoc networking (MANET): Routing protocol performance issues and evaluation considerations, Network Working Group, RFC2501, Jan 1999
3. E.C. Perkins, P. Bhagat, Highly dynamic destination-sequenced distance-vector routing (DSDV) for mobile computers, in *ACM SIGCOMM*, vol. 24, no. 4 (October 1994)
4. C.E. Perkins, E.M. Belding-Royer, S.R. Das, Ad hoc on demand distance vector (AODV) routing, IETF Internet Draft. http://www.ietf.org/internet-drafts/draft-ietf-manet-aodv-12.txt
5. D. Sudarsan, G. Jisha, A survey on various improvements of hybrid zone routing protocol in MANET. in *Proceedings of the International Conference on Advances in Computing, Communications and Informatics* (ACM), pp. 1261–1265, 3 Aug 2012

6. F. Le, G.G. Xie, H. Zhang, in On route aggregation, in *Proceedings of the Seventh Conference on Emerging Networking Experiments and Technologies, coNEXT '11* (ACM, New York, NY, USA, 2011), pp. 6:1–6:12
7. N. Beijar, Zone routing protocol (ZRP). Networking Laboratory, Helsinki University of Technology, Finland, 1–2 Apr 2002
8. Z.J. Haas, M.R. Pearlman, P. Samar, Intrazone routing protocol (IARP), IETF Internet Draft (July 2002). draft-ietf-manet-iarp-02.txt
9. Z.J. Haas, M.R. Pearlman, P. Samar, Bordercasting resolution protocol (BRP), IETF Internet Draft (July 2002). draft-ietf-manet-brp-02.txt
10. M.N. Sree Ranga Raju, J. Mungara, Optimized ZRP for MANETs and its applications. Int. J. Wirel. Mobile Netw. (IJWMN) **3**(3) (2011)
11. Z.J. Haas, M.R. Pearlman, The performance of query control schemes for the zone routing protocol. IEEE/ACM Trans. Netw. (TON) **9**(4), 427–438 (2001)
12. Z.J. Haas, M.R. Pearlman, P. Samar, The interzone routing protocol (IERP) for ad hoc networks, IETF MANET Working Group (1 June 2001). draft-ietf-manetzone-ierp-01.Txt
13. A. Subramaniam, *Power Management in Zone Routing Protocol (ZRP)* (University of Central England, Birmingham, 2003)
14. J.L. Sobrinho, L. Vanbever, F. Le, A. Sousa, J. Rexford, Scaling the internet routing system through distributed route aggregation. IEEE/ACM Trans. Netw. **PP**(99), pp. 1–15
15. Y. Wang, J. Bi, J. Wang, Towards an aggregation-aware internet routing, in *2012 21st International Conference on Computer Communications and Networks (ICCCN)* (Munich, 2012), pp. 1–7
16. F. Le, G.G. Xie, H. Zhang, Understanding route aggregation. Carnegie-Mellon Univ Pittsburgh PA School of Computer Science, 9 Mar 2010
17. Y. Rekhter, T. Li, S. Haresh, A Border gateway protocol 4 (BGP-4), RFC 4271 (Draft Standard), Updated by RFCs 6286, 6608, 6793, Jan 2006
18. F. Bohdanowicz, C. Henke, Loop detection and automated route aggregation in distance vector routing, *in 2014 IEEE Symposium on Computers and Communications (ISCC)*, (IEEE), pp. 1–6, 23 June 2014

Proposed Better Sequence Alignment for Identification of Organisms Using DNA Barcode

Sandeep Kaur, Sukhamrit Kaur and Sandeep K. Sood

Abstract DNA barcoding is a system that uses short sequence instead of whole genome; hence, it makes ecological system more accessible. It provides fast and accurate species identification. DNA barcoding, for many applications such as natural resource preservation, water quality monitoring, disease vector identification, generates short DNA sequence from standard region of genome known as marker. Methods such as BLAST, FASTA, and Smith–Waterman are generally employed for species identification using DNA barcoding. Among these methods, BLAST is used for fast species identification but gives less accurate results as compared to Smith–Waterman which is a very slow process. BLAST has been performed using a sequence to study the effect of word size on accuracy. Its results show more accuracy with smaller word size. In this study, a new algorithm with combined features of both BLAST and Smith–Waterman methods has been purposed and implemented which include more numbers of hits/matches. These hits vary with word size and threshold.

Keywords BLAST · CO1 · DNA · DNA barcode · FASTA · Identification
Marker

Abbreviation

A	Adenine
BOLD	Barcode of life data
BLAST	Basic Local Search Alignment Tool
CBOL	Consortium for barcode of life
CO1	Cytochrome c oxidase 1
CPU	Central processing unit

S. Kaur (✉) · S.K. Sood
Department of Computer Science, Guru Nanak Dev University Regional Campus,
Gurdaspur, Punjab, India
e-mail: sandeep.gndu18@gmail.com

S. Kaur
Department of Computer Science, Shanti Devi Arya Mahila College Dinanagar,
Gurdaspur, Punjab, India

© Springer Nature Singapore Pte Ltd. 2018
B. Panda et al. (eds.), *Innovations in Computational Intelligence*, Studies in
Computational Intelligence 713, https://doi.org/10.1007/978-981-10-4555-4_9

DNA	Deoxyribonucleic acid
FASTA	Fast alignment
FISH-BOL	Fish Barcode of Life
G	Guanine
Ibol	International barcode of life
ITS	Internal transcribed spacer
matK	Megakaryocyte-associated tyrosine kinase
mtDNA	Mitochondrial deoxyribonucleic acid
MUSCLE	Multiple sequence comparison by log-expectation
rRNA	Ribosomal ribonucleic acid
T	Thymine

1 Introduction to Bioinformatics

In word bioinformatics, bio means molecular and informatics is related to information technology. It means applying the informatics techniques to analyze and interpret the information related to the molecules. It is a management information system for molecular biology and can be used in medical science to identify the gene causing the disease [1]. Today, biology is playing a vital role and will be important in the coming years. Biology when depends on chemistry becomes biochemistry. When there is a need to explain biological process at atomic level, it becomes biophysics.

There is huge data collected by biologists which needs to be analyzed and interpreted using some tools of computer science resulting in new field: bioinformatics. Large storage is needed for producing large amount of data at fast rate [2, 3]. For example, growth of sequence data of GenBank is shown in Fig. 1.

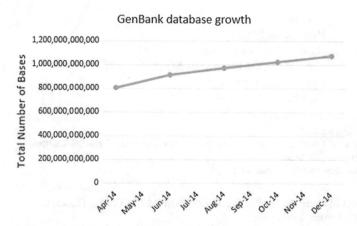

Fig. 1 Number of bases from April to December 2014 [34]

This leads to great need of information system for easy management of large amount of data. The experimental laboratory is producing over 100 gigabytes of data every day. With this large amount of data produced, improvements have been done in CPU, disk storage, and Internet for faster computations. The aims of bioinformatics are organizing the data to make it easily accessible for researchers and developing the tools for analysis and interpretation of that data [2].

Sources of data associated with the molecules can be as follows: DNA sequences made of four base letters (A, C, T, and G) of 1000 bases long, protein sequences of size 300 amino acid; macromolecular structure of size 1000 atomic coordinates; genomes of 3 billion bases; and gene expressions. The computational techniques in bioinformatics include sequence alignment, database design and data mining, phylogenetic tree (evolutionary tree) construction, functions and predicting the structure of protein, gene finding, and expression data clustering [2].

1.1 Applications of Bioinformatics

Bioinformatics has many applications in the real world in various fields such as basic research areas, medicine, microbiology, and also in agriculture. Bioinformatics helps in studying the genome comparing genomes.

Basic Research Areas

Basic research areas include functional genomics (analyzing the integration, coordination, and functions of all genes present in organism), evolutionary genomics (done by comparing the genomes of species to see how two species are relating to each other), proteomics (study of proteins), systems biology, and high-performance computing [4].

Medicine

Bioinformatics is helpful in the field of medicine such as in drug discovery, personalized medicine, and preventive medicine [4]. In drug discovery, identification of specific drugs for particular purpose is done. Personalized medicine involves the study of gene behavior of individual due to some specific medicine, i.e., how they interact with medicine. Preventive medicine involves preventing disease rather than curing it, i.e., curing the disease at early or initial stages.

Microbiology

It is the study of genomes of microorganisms to understand their behavior or environment, energy, health, and other industrial applications.

Agriculture

Bioinformatics can help in agriculture by producing better and stronger crops that can bear drought, disease by sequencing the genomes of plants [4].

2   Introduction to DNA Barcoding

Monitoring the biological effects of global climate, identification of organisms has become important to preserve species because of increasing habitat destruction. We know very less about diversity of plants and animals that are living on earth. There is an estimation of 5–50 million plants and animal species out of which less than 2 million have been identified. Yearly rate of extinction has increased from one species per million to 100–1000 species per million which means thousands of plants and animals are lost each year, most of which are not identified yet [5]. The high levels of destruction and endangerment of ecosystem have lead to improved system for identifying species. In recent years, a new ecological (preservation) approach called DNA barcoding has been proposed to identify species and ecology research [4, 6]. DNA barcoding is a system for fast and accurate species identification which will make ecological system more accessible [7]. DNA barcoding is a best tool for those who are not experts but they can do it easily using this tool even for those that are very different from each other and difficult to identify [8].

DNA barcoding first came to attention of the scientific community in 2003 when Paul Hebert's science research group at University of Guelph published a paper titled "biological identifications through DNA barcodes." DNA barcoding is a tool for identification of species and for taxonomic research. DNA barcoding is not a new concept as Carl Woese used rRNA and molecular markers such as rDNA and mtDNA to discover archaea, i.e., prokaryotes, and then for drawing evolutionary tree. But DNA barcoding uses short DNA sequence instead of whole genome for eukaryotes. Species can be identified using a short section of DNA from standard region of genome to generate DNA barcode. DNA barcode is short DNA sequence made of four nucleotide bases A (adenine), T (thymine), C (cytosine), and G (guanine). Each base is represented by a unique color in DNA barcode as shown in Fig. 1. Even nonexperts can identify species from small, damaged, or industrially processed material [9] (Fig. 2).

Identification of small damaged or processed material or dietary supplements can be done using DNA barcoding because it is a DNA-based method that uses DNA for its processing and identification, and as we know that degradation does not affect DNA of organism as it affects protein. So, because of this reason, the DNA

Fig. 2 DNA barcode [35]

Fig. 3 Evolutionary tree

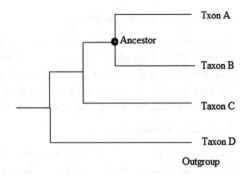

barcoding can be done using any sample even from egg to adult that means in any life stage of organisms [10].

DNA barcode should be generated from individual section of DNA. This standard or individual section also known as marker varies among the species. In animals, Paul Hebert proposed the use of CO1 or cox1 present in mitochondrial gene as marker for generating barcode, and now, it is recognized by International Barcode of Life (IBOL) as official marker for animals. The main reason for choosing mitochondrial gene is because of its small intraspecific and large inter-specific differences. But CO1 is not suitable for other group of organisms because it is uniform in them. So Internal Transcribed Spacer (ITS) is recognized for fungus, and two genes from chloroplast genome, rbcl and matK, are recognized as barcode markers for plants by IBOL [6, 11].

The sequence data generated by sequencer are used for identification and to construct a phylogenetic tree, in which related individuals are clustered together. Phylogenetic tree or evolutionary tree is a branching diagram which represents the evolutionary history of species. It can provide large amount of information. For a particular species, tree can identify ancestors and closest relatives of species. With the help of evolutionary trees many questions can be answered such as what kind of earliest animals look like or which features are inherited by their descendants [12, 13] (Fig. 3).

2.1 Applications of DNA Barcoding

1. Controlling agricultural pests

Pest damage in agriculture can cost farmers billion dollars. DNA barcoding can help with this problem by identifying pests in any stage of life which makes it easier to control them. The global Tephritid barcoding initiative contributes to management of fruit flies by providing tools to identify and stop fruit flies at border.

2. *Identifying disease vectors*

Vector species causes many serious animal and human infectious diseases like malaria. DNA barcoding allows nonecologists to identify these vector species to understand these diseases and cure them. A global mosquito barcoding initiative in building a reference barcode library can help public health officials to control these diseases causing vector species more effectively with very less use of insecticides.

3. *Sustaining natural resources*

Overharvesting of natural resources such as hardwood trees and fishes causes species extinction and economy collapse of industries that rely on them. Using DNA barcoding, natural resource managers can monitor illegal trade of products that made of these natural resources. The FISH-BOL reference barcode is a library for hardwood trees, to improve the management and conservation of natural resources.

4. *Protecting endangered species*

Primate population is reduced by 90% in Africa because of bushmeat hunting. DNA barcoding can be used by law enforcement to bushmeat in local markets which is obtained from bushmeat.

5. *Monitoring water quality*

Drinking water is a process resource for living being. By studying organism living in lakes, rivers, and streams, their health can be measured or determined. DNA barcoding is used to create a library of these species that can be difficult to identify. Barcoding can be used by environmental agencies to improve determination of quality and to create better policies which can ensure safe supply of drinking water.

6. *Routine authentication of natural health products*

Authenticity of natural health products is an important, economic, health, and conservation issue. Natural health products are often considered as safe because of their natural origin.

7. *Identification of plant leaves even if flowers or fruit are not available.*
8. *Identification of medicinal plants* [14].

2.2 Procedure of DNA Barcoding

DNA barcoding mainly have two purposes:

(a) Building the barcode library of identified species.
(b) Matching the barcode sequence of the unknown sample with the barcode library for its identification.

First of all, a specimen is collected from organism for generating the data known as DNA barcode. This specimen can be either preserved in museum or can be collected from live field. The sample then goes through laboratory processes which involve tissue sampling, DNA extraction, and sequencing. In tissue sampling, the tissue is collected from the specimen, and then, DNA is extracted from the tissue. The extracted DNA is then sequenced using sequencer. Before the sequencing of DNA, the genes are amplified polymerase chain reaction (PCR). After sequencing, a DNA sequence is generated from extracted DNA, known as DNA barcode [15]. Generally, DNA barcode is approximately 300–1000 bases in length. Here, we are not taking length in base pairs because the DNA barcode is generated in the form of only one strand of DNA, so we have considered it only in bases instead of base pairs. DNA barcode is visually represented by chromatogram. The DNA barcode can be used for two purposes: First, DNA barcode can be stored in any database for future references and to build a barcode library. For this purpose, ecologic expertise is required for selecting one or several individuals per species as reference samples in the barcode library. Second purpose is that it can be analyzed for information stored in it. This information can be used to identify the specimen for which DNA barcode is used as query sequence. This query sequence is compared with other nucleotide sequences that already exist in databases such as GenBank, BOLD. The process of comparing the sequences is known as sequence alignment. Various sequence alignment algorithms are used for sequence alignment such as pairwise sequence alignment algorithms and multiple sequence alignment. In my work, I have focused on pairwise sequence alignment and some of its important algorithms [15] (Figs. 4 and 5).

2.3   Users of DNA Barcoding

As we know that DNA barcoding is a tool for identifying and classifying species and to create an online database containing DNA sequences of various organisms that can be used as future reference. If DNA sequence extracted from the sample is matched with any sequence of database, then sample is identified; otherwise, it is a

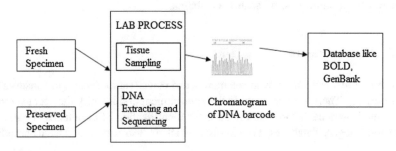

Fig. 4 DNA barcoding procedure

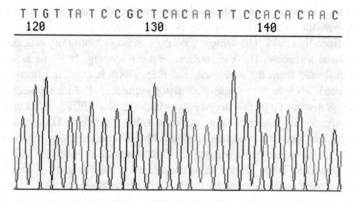

Fig. 5 Chromatogram of DNA barcode generated by sequencer [36]

new species that is not yet discovered or identified. DNA barcoding is a tool that is very helpful for users for different purposes.

2.3.1 Taxonomist

Taxonomist is a biologist that studies the different organisms and categorizes them according to the relationship between them. So, DNA barcoding is very helpful for taxonomists as finding the relationship becomes very easy with it. A taxonomist can use DNA barcoding for studying the nature and its distributions for which he or she must have good skills and can gather information from literature, museum, or databases [16].

2.3.2 Ecologists

Ecologists study the environment, organisms living in that environment, and how actions of human affect other living organisms. DNA barcoding is a great tool to identify the nature and its distribution. Information needed for this purpose can be gathered from literatures, museums, databases.

2.3.3 Conservationist

Conservationist is a person who is a member of conservation system to conserve the environment. Their work is to study the environment and identify the species living in the environment. Information needed for identification of organisms is gathered from the images, databases, taxonomists which is also a user of DNA barcoding [16].

2.3.4 Legal Users

Legal users are like police. These users need identification for the purpose of forensic investigation, controlling the crimes related to wildlife, and illegal trades. Information needed for this purpose can be gathered from images, databases, and taxonomists.

2.3.5 Animal/Human Health

The users that deal with animal and human health also use the benefits of DNA barcoding. These are the persons who find the medicinal and harmful properties in the organisms because some animals can be very helpful for other organisms and in making good medicines. The sources of information needed for this purpose are images, databases, and taxonomists. Their skills and interest are medium, i.e., varying [16].

2.3.6 Environmental Protection

The user that deals with environment protection is the persons that do the identification of invasive species and indicator species. Invasive species are those which are dangerous to other organisms in environment and have harmful effects on them, e.g., pests that have harmful and dangerous effects on the species on which they are living. For example, fungus and bacteria are pests. The indicator species are those species whose presence in an area indicates some environmental effects on other species in the same environment. Like, due to some changes or conditions occurring in that area, scientists can examine the indicator species, to check that how these changes or conditions have affected the other species that are very difficult to examine. The information needed for this purpose can be gathered from the images, databases, and taxonomists. Their skills and interest are not good and not bad.

2.3.7 Naturalist

Naturalists are the users that deal with the environment. They identify the organisms living in the environment to study the environment and its classification and must have good skills. The information needed for this purpose can be collected from literature, databases, and taxonomists.

2.3.8 Users that Deal with Utilization of Biodiversity

These users deal with the identification of plants, crops, or animals for the purpose of their utilization, so that they can be utilized by other organisms in the environment. The information needed for this purpose can be gathered from images, field guides, taxonomists, and online databases.

2.3.9 Public

These users are people from public. Some of them may be interested and other may be uninterested. They do identification due to some occasional purpose like some people want to study for their thesis or other research work. The source of information needed for identification is images or guides [16].

2.4   Future of DNA Barcoding

DNA barcoding is like a revolution in history of identification of organisms. DNA-based identification has linked the various samples collected from the field or museum with the sequences present in the online databases such as BOLD (Barcode of Life Data) and GenBank. Various questions have asked by many biologists about its validity and its benefits in the future. As long as there is a need for conservation of nature and natural resources and its right utilization, there is a need of identification of organisms [16]. DNA barcoding is a much better identification tool than the traditional ones and previous methods. DNA barcoding has been succeeded for distinguishing and identifying the organisms living in the environment. Also it has been helpful in case of cryptic species which are closely related to each other or looks same but slight biological difference. DNA barcoding is fast and more accurate tool which can be used to build a device to identify the species based upon the input sample with in few minutes or seconds [17].

2.5   Limitation in DNA Barcoding

The aim of DNA barcoding is identification of samples and species discovery. Identification of specimen involves the process of naming the unidentified specimen using DNA barcoding. There are many benefits of DNA barcoding but have some limitations too. [18] Accuracy of diagnosis of an organism is dependent on the intraspecific variation which is compared with intraspecific differences of other species. With this using a short DNA sequence, we can check that in which taxon a particular species or organism fits. This decision should be made with confidence [19]. Some limitations of DNA barcoding are given below.

2.5.1 Unclear Objective Hypothesis

The main aim of DNA barcoding is to build a DNA barcode reference library, then validating and testing the library for future use and last exploring the unidentified organisms. Validation and testing of reference library and exploring the unidentified species are mixed because the reference library should be tested and must be corrected and validated. The identification of organisms depends on the data of reference library. So, if the reference library is not tested, then it will affect the identification.

2.5.2 Incorrect Identification of Samples

The second limitation is the human error and inaccuracies made while creating the reference library. These errors can become big problems so reference library should be tested. With this, there can be conflicts between identification of same species by different persons. Like when two different persons are working in different laboratories but on same taxa, then the errors made by them in reference library can result in different identification for same taxa. So, there is a need for correct maintenance of reference library data.

2.5.3 Interpretation of Species Identification

The term species identification causes a big confusion in species discovery and sample identification. The actual meaning of species identification is identification of sample to species. To avoid or minimize the confusion, the concept of species discovery and sample identification should be clearly defined [18].

3 Literature Survey

Luscombe [2] hase explained the need of computer resources in bioinformatics. It has been explained that large amount of data and information is produced which needs to be handled, managed, analyzed, and interpreted. This can be done by using tools available in bioinformatics. Bioinformatics employs number of computational techniques such as sequence alignment, data mining, evolutionary tree generating, genome sequences, gene findings, and prediction of protein structure.

Hebert et al. [20] have stated that identification of organism lies in the construction of systems that uses DNA sequences as barcodes. In this paper, author has established that mitochondrial gene cytochrome c oxidase 1 (CO1) can serve as main part of identification system for animals. CO1 has taken from mitochondrial gene of genome in animals and has 100% successful in identification of animals. CO1 identification system will provide a reliable, accessible, and cost-effective solution to the current problem of species identification.

Cohen [3] has explored the importance of computer science in bioinformatics. In this paper, author has introduced computer scientists with the new field bioinformatics. Bioinformatics has born with the needs of biologists for collecting, managing, analyzing, and interpreting data. This all can be done with resources available in computer science such as hard disk, CPU, and Internet.

Cohen [3] explained the importance of information technology in biology or medical field. Nowadays, there are lot of experiments going on in many laboratories in the world such as exploring the genes, their function, medicine field and their effect on genes, study of human genome, and genomes of other organisms. These experiments are carried out by biologists. In these experiments, huge amount of data is produced which needs to be handled, needs to be stored somewhere, and needs to be processed. For these purposes, there is a need of information technology and its resources such as storage, CPU, computing power in biology. Biologists develop some algorithms for operations to be done using information technology resources. Use of information technology in biology is known as bioinformatics. Bioinformatics reduces the complexity of these experiments and operations which are to be done in these experiments. The author has described the software developed for biologists and currently being used by them and some areas of computer science that is very important in field of bioinformatics.

Hebert and Gregory [7] have discussed that DNA barcoding is an appropriate system developed to provide fast, accurate, and automatable species identification by using short DNA sequence generated from standardized region of gene. It makes taxonomic system more accessible, beneficial to ecologists and conservationists. One day, DNA barcoding will lead us to a state when everyone will have easy access to name and biological attributes of any species on the planet. It also highlights diversity in species and may represent new species. Even though it is beneficial, it also has been controversial in some scientific areas.

Notredame [21] has discussed about multiple sequence alignment and its various algorithms. These algorithms are based on different scoring schemes such as matrix-based scoring schemes that include algorithms like ClustalW and consistency-based scoring schemes that include algorithms like T-Coffee. Then, the procedures of these algorithms are also discussed.

Little et al. [22] have concluded that to use DNA sequences for species identification, an algorithm to compare the sequences is needed. Two novel alignment-free algorithms were used to identify query sequences for the purpose of DNA barcoding. Gymnosperm nrITS2 and plastid matK sequences were used on test data. Results show that DNA barcoding could be used to identify samples with a very less error. Geographic range can be used as elimination factor without which DNA barcoding does not appear to be useful for species-level identification.

Waugh [23] discussed that since 250 years lot of work in systematic has done but majority of species are still unidentified. Increase in extinction rates and need of biological monitoring lead to DNA barcoding. So DNA barcoding is a technology that has been proposed that might expedite the species identification. This method involves various markers for various species for efficient results. For this, there is a need of various identification programmers to be coordinate. DNA barcoding may prove to be very useful taxonomic tool.

Vaidya et al. [1] have stated that cancer is widely spreading worldwide. One of the types of cancer is breast cancer. In this paper, bioinformatics and its process are explained. It includes that study of breast cancer involves predicting its causes, drugs for its cure, and predicting its transmission in the next generations.

Dalton et al. [24] have discussed that smuggling of wildlife animals for commercial purpose has lead to population decline in South Africa. So, mitochondrial CO1 gene was sequenced to determine species of unknown sample in three suspects of South African forensic wildlife cases. Two unknown samples were identified as domestic cattle and third was identified as common reedbuck.

Kress and Erickson [15] have stated that DNA barcoding is a new method for fast identification of species used in DNA based on DNA sequences, generated from small standardized region of genome. As a research tool for ecologists, it expands ability to diagnose species by including all life-history stages of organisms. The DNA barcoding involves building of DNA barcode library of known species and then matching barcode sequence of unknown sample against the barcode library for identification. It has grown because number of sequences has generated as barcode and in terms of its application.

Nagy et al. [25] have concluded that DNA barcoding is now a popular and well-accepted tool for identification of various species and detection of taxonomic diversity, since its introduction in 2003. This method is becoming an essential part of taxonomic practice. DNA barcoding is a tool that makes species identification easier and also bodies such as iBOL (Barcode of Life Project) and CBOL (Consortium for the Barcode of life) for their projects.

Abilash and Rohitaktha [26] conclude that pairwise sequence alignment is one of the methods to arrange two biological sequences to identify similarity which indicates functional, structural relationship between them. Pairwise alignment has two methods: local and global alignment methods. Local alignment is applicable in searching local similarity in large sequences. Global alignment aligns the sequences by taking whole sequence at a time. Local and global pairwise alignment methods are analyzed to find out similarity between the sequences.

Khallaf et al. [27] stated that due to substitution of species, accurate identification of seafood species in markets is a growing concern. It has become prime priority of governments to identify the already processed fish products. DNA barcoding was applied to some samples purchased from Egyptian markets and were analyzed. Sequencing of mitochondrial cytochrome c oxidase (CO1) gene revealed 33.3% species substitution in fish products which demonstrates that DNA barcoding is a reliable tool for detecting fish products.

4 Problem Definition

1. Alignment of DNA sequences for checking similarity using small word size and gaps.
2. Comparison of various sequence alignment algorithms.

4.1 Sequence Alignment

Sequence alignment is a process of comparing two or more sequences, whether DNA, RNA, or protein sequence, to look for similar patterns in sequences [13, 28]. DNA sequence is made of four bases A (adenine), T (thymine), C (cytosine), and G (guanine), and for identification of species, these need to be aligned. Comparison of sequences has become very helpful in understanding the information content and functions of genetic sequence and can tell that how much the sequences are closely related. Sequence alignment provides solution to many problems in bioinformatics including identification of new species, finding relationship between species, and for predicting the function and structure of genes and proteins [29]. Sequence alignment is a process of looking for similarity regions in two sequences. Aligning two sequences is done to get a most suitable alignment. Suitable alignment means aligning the two sequences in such a way that it gives maximum score to do alignment. The alignment is done to look for similarity in two sequences. If there are more matches in sequences, then sequences are more similar, and if less matches, then less similar. While aligning the sequences, there are three possibilities: gap or indel (means insertion deletion), match, and mismatch.

Let us take an example of very small sequences:

$$S1 = ATCG$$
$$S2 = TCA$$

We can align these sequences in various ways; with each alignment, we will get different scores, but we will consider only that which alignment gives us maximum score value. Let us suppose the values have gap, match, and mismatch. The value of match must be positive and value of gap and mismatch must be negative.

$$\text{Let value of Match} = 2$$
$$\text{Mismatch} = -1$$
$$\text{Gap} = -2.$$

First way to align is as following:

S1	A	T	C	G
S2		T	C	A

```
      Gap  match match mismatch
      -2    +2    +2    -1
                           Score= -2+2+2-1 = 1
```

Second way to align is as following:

S1	A	T	C	G
S2	T	C	A	

```
   Mismatch mismatch mismatch  Gap
      -1       -1       -1     -2
                     Score= -1-1-1-2 = -5
```

Third way to align is as following:

S1	A	T	C	G
S2	T		C	A

```
   Mismatch  gap    match  mismatch
      -1     -2      +1      -1       S                          Score= -1-2+1-1 = -3
```

From the above three ways, first method gives us maximum score, hence is a suitable alignment. The alignment can be done only of those sequences which are having some similarity [28].

DNA sequence alignment is of two types:

(a) Pairwise sequence alignment and
(b) Multiple sequence alignment.

(i) *Pairwise Sequence Alignment*

Pairwise sequence alignment is a method of aligning two sequences at a time. The sequences can be of same or different size or they can be somewhat similar or dissimilar sequences.

(ii) *Multiple Sequence Alignment*

Multiple sequence alignment is a process of aligning or comparing more than two or three sequences with database sequences. This is done to create evolutionary tree or phylogenetic tree and to predict the structures like predicting the structure of proteins. With multiple sequence alignment algorithms, the study and analysis of relationship between various taxa using phylogenetic or evolutionary tree has become easy. The accurate multiple sequence alignment is not possible but some heuristic algorithms are used for aligning the multiple sequence. There are two scoring techniques in multiple sequence alignment and that is matrix-based scoring scheme and consistency-based scoring scheme. The multiple sequence alignment algorithms with matrix-based scoring scheme are ClustalW, MUSCLE, and Kalign, and algorithms with consistency-based scoring scheme is T-Coffee [21].

4.2   Already Existing Pairwise Alignment Algorithms

Pairwise alignment is a process of aligning two sequences at one time to check for similarity between them. These methods are used to find the best matching local or global alignments of two sequences. For example, if two sequences are taken from different organisms and from a common ancestor, then because of similarity, they will get aligned. The purpose of this arrangement is to determine the relationship between the biological sequences [26]. It is based on a score which is evaluated from the number of same characters in two sequences, number and length of gaps, required to align sequence so that the two sequences get aligned [30]. Alignments can be of two types: local alignment and global alignment. Global alignment technique involves the attempt to align every character in every sequence. In this, number of characters in sequences or size should be same. This approach would be time-consuming and inconvenient for longer sequences. In global alignment, the two sequences of equal length are aligned completely or globally. Full sequence is aligned to look for similarity. Local alignments are appropriate for dissimilar sequences which may contain similar character sequence. In local alignment, the two sequences that can be of equal or different length and of similar or dissimilar are aligned to look for similarity in local regions. It means two sequences are compared, and whereever similarity occurs, that local region is counted as having similarity [26].

4.2.1   Needleman–Wunsch Algorithm

The Needleman–Wunsch algorithm was published in 1970. It performs a global alignment on two nucleotide or protein sequences. This algorithm provides a method of finding the ideal global alignment of two sequences by maximizing the matches and minimizing the number of gaps that are necessary to align the two sequences. The alignment with the highest score must be the best alignment for which score matrix has to be prepared. Algorithm is as follows.

A and B are sequences and Ai and Bj represent the bases of sequence at position i and j.

Step 1: Score matrix is created.
Step 2: Trace backing is done.
Step 3: Compute an alignment that actually gives this score, start from the bottom right cell, and compare the value with the three possible sources (diagonal, up, and bottom) to see which it came from. If diagonal, then Ai and Bj are aligned, if up, then Ai is aligned with a gap, and if left, then Bj is aligned with a gap.

To analyze the time complexity of the Needleman–Wunsch algorithm, we have to analyze each part of the algorithm. For start filling the score matrix, time complexity of $O(M + N)$. The next step is filling other cells of the matrix with all the scores. For each cell of the matrix, three neighboring cells must be compared, which needs constant time. So, to fill the entire matrix, the time complexity is the number of entries, i.e., $O(MN)$. Finally, the trace back requires a number of steps. We can move a maximum of N rows and M columns, and thus, the complexity of this is $O(M + N)$. The second step is finding the final path which involves jumping from cells of matching characters. Since this step can include a maximum of N cells, this step is $O(N)$.

Thus, the overall time complexity of this algorithm is

$$O(M+N)+O(MN)+O(M+N)+O(N) = O(MN)$$

Since this algorithm fills the cells of single matrix of size MN, i.e., $O(MN)$, and stores at most N positions for the trace back, i.e., $O(N)$, the total space complexity of this algorithm is $O(MN)+O(N) = O(MN)$.

(i) *Methodology*

Needleman–Wunsch algorithm is performed using two nucleotide sequences of size 368 base pairs. These sequences are collected from NCBI database of nucleotides.

Two sequences are taken as inputs. Source of query sequence is Homo sapiens and the sequence is as below:

ACAAGATGCCATTGTCCCCCGGCCTCCTGCTGCTGCTGCTCTCCGGG
GCCACGGCCACCGCTGCCCTGCCCCTGGAGGGTGGCCCCACCGGCCGA
GACAGCGAGCATATGCAGGAAGCGGCAGGAATAAGGAAAAGCAGCCT
CCTGACTTTCCTCGCTTGGTGGTTTGAGTGGACCTCCCAGGCCAGTGCC
GGGCCCCTCATAGGAGAGGAAGCTCGGGAGGTGGCCAGGCGGCAGGA
AGGCGCACCCCCCCAGCAATCCGCGCGCCGGGACAGAATGCCCTGCA
GGAACTTCTTCTGGAAGACCTTCTCCTCCTGCAAATAAAACCTCACCC
ATGAATGCTCACGCAAGTTTAATTACAGACCTGAA [31].

Source of other sequence is Tasmanian Mountain Shrimp, and the sequence is as below:

TCTTTAGATTTTATTTTTGGAGCTTGGTCTGGCATAGTAGGCACCGCC
CTAAGACTTATTATTCGGGCTGAATTAGGACAACCTGGTAGACTTATTG
TGATGATCAAATTTACAACGTGGTCGTAACAGCTCATGCTTTTGTGATA
TTTTTTTTTATAGTTATGCCCATTATAATTGGTGGATTTGGAAATTGACTT
TTCCCTTAATATTAGGTGCTCCTGATATAGCTTTTCCTCGTATAAATAAT
ATAAGATTTTGACTTCTTCCACCTTCTTTAACTCTTCTCCTATCCAGAGG
AATAGTTGAAAGAGGTGTTGGCACAGGATGAACTGTTTATCCTCCTTTA
GCTGCTGGAATCGCCCATGC [31] (Fig. 6).

Needleman–Wunsch algorithm has been performed on Web site: http://www.ebi.ac.uk/Tools/psa/. This Web site provides the pairwise sequence alignment algorithms such as Needleman–Wunsch or Smith–Waterman algorithm. The score of this alignment is 197.5.

4.2.2 Smith–Waterman Algorithm

The Smith–Waterman algorithm is a well-known algorithm used for local sequence alignment, i.e., for determining similar characters or patterns between two

```
AB000263       1 -----ACAAGATGCCATTGTCCCCCGGCCTCCTGCTGCTGCTGCTCTCCG   45
                    ||||...|||.|              ||..|||  ||.|||
EMBOSS_001     1 TCTTT---AGATTTTATTTT-----------TGGAGCT--TGGTCT---   30

AB000263      46 GGGCCACG--GCCACCGCTGCCCTGCCCCTGGAGGGTGGCCCCACCGGCC   93
                    |||...|  |.||||   |||||
EMBOSS_001    31 -GGCATAGTAGGCACC---GCCCT-------------------------   50

AB000263      94 GAGACAGCGAGCATATGCAGGAAGCGGCAGGAATAAGGAAAAGCAGCCTC   143
                    .|||        |.|||    .|..|||...||||.||||.||      |
EMBOSS_001    51 AAGA-------CTTAT----TATTCGGGCTGAATTAGGACAA-------C   82

AB000263     144 CT----GACTTTCCTCGCTTGGTGGT-------TTGA----GTGGACCTC   178
                    ||    |||||      .||||||.|       ||.|    ||||.|.|.
EMBOSS_001    83 CTGGTAGACTT------ATTGGTGATGATCAAATTTACAACGTGGTCGTA   126

AB000263     179 CCAGGCCAGTGC-----------------------CGGGCCCCTCATAG   204
                    .|||..|| |||                       ...||||.|.|||
EMBOSS_001   127 ACAGCTCA-TGCTTTTGTGATAATTTTTTTTTATAGTTATGCCCATTATA-   174

AB000263     205 GAGAGGAAGCTCGGGA--GGTGGCCAGGCGGCA---GGAAGGCGCACCCC   249
                    |  |||||        |  |||
EMBOSS_001   175 ---------------ATTGGTGG---------ATTTGGA-----------   189

AB000263     250 CCCAGCAATCCGCGCGCCGGGACAGAATG--CCCT-----GCAGGAACTT   292
                    |||        .|||  .|| ||||      ..|||    |.
EMBOSS_001   190 ------AAT----------TGAC---TTGTTCCCTTAATATTAGG---TG   217

AB000263     293 CTTCTGGAAGACCTTCTCCTCCTGCAAATAAAA----------------   325
                    ||.|||..|.|.|||.|||||.|...|||||||.|
EMBOSS_001   218 CTCCTGATATAGCTTTTCCTCGTATAAATAATATAAGATTTTGACTTCTT   267

AB000263     326 ----------------CCTCACCCAT------GAATGCTCACGCAAGTTTA   354
                    |.||.||.||       ||||         ||||.|
EMBOSS_001   268 CCACCTTCTTTAACTCTTCTCCTATCCAGAGGAAT---------AGTTGA   308

AB000263     355 A-------TT---ACA-GACCTGAA------------------------   368
                    |       ||   ||| || ||||
EMBOSS_001   309 AAGAGGTGTTGGCACAGGA--TGAACTGTTTATCCTCCTTTAGCTGCTGG   356

AB000263     369 ------------   368

EMBOSS_001   357 AATCGCCCATGC   368
```

Fig. 6 Output of Needleman–Wunsch algorithm [32]

nucleotide or protein sequences. Instead of looking at the total sequence, the Smith–Waterman algorithm compares segments of all possible lengths and checks for similarity.

The Smith–Waterman algorithm was published in 1981 and is very similar to the Needleman–Wunsch algorithm. But still it is different because it is a local sequence alignment algorithm. Instead of aligning the entire length of two sequences, it finds the local region of highest similarity.

Step 1: Score matrix is created. All cells have values either 0 or 1.

Step 2: Trace backing is done. It starts with the maximum value in score matrix.

Step 3: Now compute the alignment, the local alignment value takes the maximum value of all the three values taken in the global alignment with the value "0". And trace back starts with the maximum value in the score matrix and traverse diagonally aligning every character of both the sequences until it encounters the value "0" in the score matrix [26].

To analyze the time complexity of the Needleman–Wunsch algorithm, we have to analyze each part of the algorithm. For start filling the score matrix, time complexity of $O(M + N)$. The next step is filling other cells of the matrix with all the scores. For each cell of the matrix, three neighboring cells must be compared, which need constant time. So, to fill the entire matrix, the time complexity is the number of entries, i.e., $O(MN)$. The time complexity for the trace back is $O(MN)$. The time complexity of Smith–Waterman algorithm is same as Needleman–Wunsch algorithm.

Therefore, the total time complexity of the Smith–Waterman algorithm is

$$O(M+N) + O(MN) + O(MN) = O(MN)$$

The space complexity of the Smith–Waterman algorithm is also same as the complexity of Needleman–Wunsch algorithm. This is because same matrix is used and same amount of space for trace back is needed.

(i) *Methodology*

Smith–Waterman algorithm is performed using two nucleotide sequences of size 368 base pairs. These sequences are collected from nucleotide database of NCBI. In this algorithm, same nucleotide sequences are used that are used in Needleman–Wunsch algorithm.

Smith–Waterman algorithm has been performed on Web site: http://www.ebi.ac.uk/Tools/psa/. This Web site provides the pairwise sequence alignment algorithms such as Needleman–Wunsch or Smith–Waterman algorithm. The score of this alignment is 209.0 (Fig. 7).

4.2.3 FASTA

FASTA stands for fast alignment. FASTA is a fast searching algorithm used for comparing query sequence with database. It comes under dynamic programming and it was developed by Lipman and Pearson in 1985. FASTA is faster than Smith–Waterman and Needleman–Wunsch algorithms which are good for two-sequence comparison, but when to compare with entire database, they are very slow than FASTA. The algorithm is as follows:

```
AB000263     73 TGGAGGGTGGCC-----------CCACCGGCCGAGACAGCGAG-CATATG    110
                ||||| ..|||.|           .||||| .||.|      || |.|||
EMBOSS_001   18 TGGAGCTTGGTCTGGCATAGTAGGCACCGCCCTA-------AGACTTAT-    59

AB000263    111 CAGGAAGCGGCAGGAATAAGGAAAAGCAGCCTCCT-GACTTTCCTCGCTT   159
                .|..|||...||||.||||    ||.|||..| |||||     .||
EMBOSS_001   60 ---TATTCGGGCTGAATTAGGA----CAACCTGGTAGACTT------ATT   96

AB000263    160 GGTGGT-------TTGA----GTGGACCTCCCAGGCCAGTGCCGGGCCCC   198
                |||||.|      ||.|   ||||.|.|..|||                |
EMBOSS_001   97 GGTGATGATCAAATTTACAACGTGGTCGTAACAG---------------C   131

AB000263    199 TCATAGGAGAGGAAGCTCGGGAG----------------GTGGCCA--   228
                ||||        |||...|.|            .||.|||
EMBOSS_001  132 TCAT---------GCTTTTGTGATAATTTTTTTTATAGTTATGCCCATT   171

AB000263    229 ------GGCGGCA---GGAAGGCGCACCCCCCCAGCAATCCGCGCGCCGG   269
                   ||.|| |   |||                   |||          .
EMBOSS_001  172 ATAATTGGTGG-ATTTGGA----------------AAT----------T   193

AB000263    270 GACAGAATG--CCCT-----GCAGGAACTTCTTCTGGAAGACCTTCTCCT   312
                |||   .|| ||||     ..|||  |.||.|||..|.|.|||.||||
EMBOSS_001  194 GAC---TTGTTCCCTTAATATTAGG---TGCTCCTGATATAGCTTTTCCT   237

AB000263    313 CCTGCAAATAAAA------------------------------CCTCA   330
                |.|..||||||.|                               |.||.
EMBOSS_001  238 CGTATAAATAATATAAGATTTTGACTTCTTCCACCTTCTTTAACTCTTCT   287

AB000263    331 CCCAT------GAATGCTCACGCAAGTTTAA-------TT---ACA----   360
                ||.||      ||||     |||| .||      ||   |||
EMBOSS_001  288 CCTATCCAGAGGAAT---------AGTTGAAAGAGGTGTTGGCACAGGAT   328

AB000263    361 GACCTG     366
                ||.|||
EMBOSS_001  329 GAACTG     334
```

Fig. 7 Output of Smith–Waterman algorithm [32]

I is query sequence and *J* is test sequence.

Step 1: Identify common *k* words or simply words between *I* and *J* using a dot plot matrix. For DNA *k* = 6, i.e., 6 nucleotides.
Step 2: Score diagonals with *k* word matches, identify 10 best diagonals.
Step 3: Rescore initial region with a substitution score matrix.
Step 4: Join initial regions for gaps.
Step 5: Perform dynamic programming for final alignment.

The complexity of the FASTA algorithm depends on the size of the *k-tuples*, which means the larger the *k-tuples*, the faster the algorithm. The true complexity is not easily determined because the speed of the alignment of two sequences depends on the total number of marked cells' variable diagonals. The space complexity of this algorithm is also *O(MN)* like the Needleman–Wunsch and Smith–Waterman because it uses a matrix. But it uses less space because not all cells in the matrix are marked.

4.2.4 BLAST

BLAST stands for Basic Local Alignment Search Tool. TBLAST algorithm was developed by Altschul, Gish, Miller, Myers, and Lipman in 1990 to increase the speed of FASTA by finding fewer and better spots of denser matching during the algorithm. BLAST concentrates on finding local regions of high similarity in alignments without gaps.

Algorithm:

Step 1: Using word search method, sequence is filtered to remove complexity regions.

Step 2: Identification of exact word match method searches the database for neighborhood word. Words having equal or greater scores than neighborhood score threshold are taken for alignment.

Step 3: Using maximum segment pair alignment method, it extends the possible match as ungapped alignment in both directions that stops at maximum score.

The complexity of the BLAST algorithm is *O(MN)*. This is the same time complexity as all of the other algorithms, but BLAST significantly reduces the numbers of segments which need to be extended so the algorithm runs faster than all the previous algorithms. Using BLAST for nucleotide sequences, DNA barcoding has been used as a tool for identification of three species in forensic wildlife in South Africa [24] and also it has revealed high level of mislabeling in fish fillets purchased from Egyptian markets [27].

(i) *Methodology*

BLAST has been performed using two nucleotide sequences of size 368 base pairs. These sequences are collected from nucleotide database of NCBI. In this algorithm, same nucleotide sequences are used that are used in Needleman–Wunsch algorithm and Smith–Waterman algorithm (Fig. 8).

BLAST has been performed on Web site: https://blast.ncbi.nlm.nih.gov/. There are numbers of alignments in the output and each alignment has different score (Table 1).

🖥Download ∨ Graphics Sort by: [E value   ▼]

EMBOSS_001

Sequence ID: lcl|Query_37791 Length: 368 Number of Matches: 16

Range 1: 229 to 237 Graphics   ▼ Next Match ▲ Previous Match

Score	Expect	Identities	Gaps	Strand
17.5 bits(18)	0.68	9/9(100%)	0/9(0%)	Plus/Minus

```
Query   129   AGGAAAAGC   137
              |||||||||
Sbjct   237   AGGAAAAGC   229
```

Range 2: 6 to 16 Graphics   ▼ Next Match ▲ Previous Match ▲ First Match

Score	Expect	Identities	Gaps	Strand
15.7 bits(16)	2.4	10/11(91%)	0/11(0%)	Plus/Minus

```
Query   318   AAATAAAACCT   328
              |||||||| ||
Sbjct   16    AAATAAAATCT   6
```

Range 3: 232 to 239 Graphics   ▼ Next Match ▲ Previous Match ▲ First Match

Score	Expect	Identities	Gaps	Strand
15.7 bits(16)	2.4	8/8(100%)	0/8(0%)	Plus/Plus

```
Query   149   TTTCCTCG   156
              ||||||||
Sbjct   232   TTTCCTCG   239
```

Range 4: 261 to 268 Graphics   ▼ Next Match ▲ Previous Match ▲ First Match

Score	Expect	Identities	Gaps	Strand
15.7 bits(16)	2.4	8/8(100%)	0/8(0%)	Plus/Plus

```
Query   289   ACTTCTTC   296
              ||||||||
Sbjct   261   ACTTCTTC   268
```

Fig. 8 Output of BLAST [33]

Table 1 Comparison of sequence alignment algorithms [26]

	Complexity	Alignment	Accuracy	Speed
Needleman–Wunsch	$O(MN)$	Global alignment	Less accurate than Smith–Waterman	Slow for searching entire database
Smith–Waterman	$O(MN)$	Local alignment	More accurate	Slow for searching entire database
FASTA	Time complexity depends on k	Local alignment	Less accurate than Smith–Waterman	Faster than above
BLAST	$O(MN)$	Local alignment	Less than Smith–Waterman	Fastest

5 Proposed Work

Proposed work of thesis is to do sequence alignment with more accuracy and this idea has been implemented with combining the features of Smith–Waterman and Basic Local Alignment Search Tool (BLAST).

In the proposed model of sequence alignment algorithm, the concept of gapped alignment from Smith–Waterman algorithm is combined with the word size and heuristic approach of BLAST algorithm. In this model, first of all, query sequence is broken into words of size 3, 4, or 5. The small size of words is to get more numbers of hits while matching because with small word the small matches cannot be missed. Then these words are stored in indexed table. Suppose we have query sequence ACTGACTGCCCGTAAATGCATCGTAGC. Now with word size $006B = 3$, underlined words are stored in the table with their indices as shown below.

ACTGACTGCCCGTAAATGCATCGTAGC

Then from the indexed table, the words are matched with sequences present in the database. Then, these words are matched with query database and aligned with insertions and deletions. Then, theses aligned words are extended to both left and right directions till the score is increasing. Finally, the highest scored pairs are chosen (Fig. 9).

Fig. 9 Proposed work

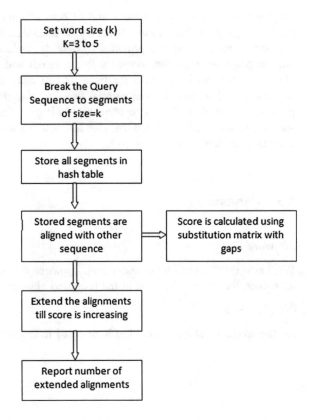

5.1   Proposed Algorithm

Step 1: Decompose the query sequence into words of length k, use $k = 3$ to 5.
Step 2: Store all words in hash table for faster searching and matching.
Step 3: For each word, look into hash table with a score greater than threshold. The scores are calculated using a substitution matrix by including gaps in sequences. These gaps are also known as indel (insertions and deletions).
While comparing sequences A and B, if gap is inserted in B then it is known as deletion and in sequence A, at corresponding base it will be insertion.
Step 4: Search the database for sequences containing any one of the words.
Step 5: Extend the hit (matched word) in both directions until its score is increasing.
Step 6: Report the highest scoring pairs if its score is greater than the threshold.

5.2   Features in Proposed Algorithm

Proposed algorithm is the combination of best features of Smith–Waterman and BLAST algorithm. That means accuracy of identification is provided by Smith–Waterman and fast search is provided by heuristic technique of BLAST algorithm. So, the proposed algorithm provides faster search and accurate results. Also the word size used will be 3 to 5 for faster search and more sensitivity (accuracy). Because speed is directly proportional to word size and sensitivity is inversely proportional to word size large word size will give faster search speed and less sensitivity, and small word size will give less search speed and more sensitivity. So, the word size has chosen to be 3–5.

5.3   Parameters

(a) *Word Size*

Word size is the size of word taken from sequence that is used for searching in the databases. Its value to be used in the proposed algorithm will be 3–5.

(b) *Threshold*

All the words must have scored at least equal to threshold.

5.4 Methodology

For implementing the proposed work, first we have to check whether sensitivity increases with reduction in word size or not. For this, BLAST has been performed with different word size and then number of hits or number of matches is recorded. Parameters to be considered are word length or word size and number of hits in the sequence alignment. Word size denoted by k is the length of word or segment that is to be used as size of segment of sequence before starting the alignment. Number of hits is number of matches or alignments found in the sequences. To study the effect of change in work length on accuracy of species identification, BLASTN was performed using nucleotide sequence of invertebrate animal species named as Anaspides tasmaniae against nonredundant database. It returns top 100 sequences having some similarity, for each query sequence [1]. It is a freshwater species, i.e., common resident of lakes, streams, and pools in caves, in Tasmania highlands. To observe the effect of word length parameter, values of 7, 11, and 15 are used with the expect value, $E = 10$.

The sequence used for the observation which is extracted from CO1 region of Anaspides species of size 657 bases, and in FASTA format is as follows. This sequence is collected from NCBI database of nucleotides.

>EMBOSS_001

TCTTTAGATTTTATTTTTGGAGCTTGGTCTGGCATAGTAGGCACCGCC
CTAAGACTTATTATTCGGGCTGAATTAGGACAACCTGGTAGACTTATTG
GTGATGATCAAATTTACAACGTGGTCGTAACAGCTCATGCTTTTGTGAT
AATTTTTTTTATAGTTATGCCCATTATAATTGGTGGATTTGGAAATTGAC
TTGTTCCCTTAATATTAGGTGCTCCTGATATAGCTTTTCCTCGTATAAAT
AATATAAGATTTTGACTTCTTCCACCTTCTTTAACTCTTCTCCTATCCAG
AGGAATAGTTGAAAGAGGTGTTGGCACAGGATGAACTGTTTATCCTCC
TTTAGCTGCTGGAATCGCCCATGCAGGCGCTTCTGTGGACTTAGGAATT
TTTTCTCTTCATATAGCGGGAGCTTCTTCTATTTTAGGGGCGGTAAATTT
TATTACTACTTCTATTAATATGCGTGCCAATGGTATAACTTTAGATCGA
ATACCTTTATTTGTCTGATCCGTTTTTATTACTGCTATTCTTTTACTACTC
TCTCTTCCCGTTTTAGCAGGGGCAATCACAATACTTCTCACTGACCGTA
ACTTAAATACTTCTTTCTTTGACCCCGCTGGAGGAGGAGATCCATTCTT
TATCAACATAAATGCC [32] (Table 2).

Table 2 Varying number of hits with different word size	Word size (k)	No. of hits
	7	518,310,295
	11	32,757,086
	15	14,769,504

Fig. 10 Comparison on the
basis of word size

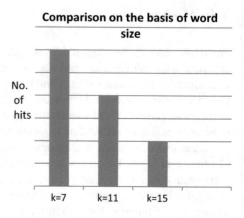

The results from word size $k = 7$ returned 518,310,295 hits, $k = 11$ returned 32,757,086, i.e., less hits than returned by $k = 7$ and then $k = 15$ returned 14,769,504 which is least of all. So the observations tell us that decreasing the word size gives more number of hits, i.e., more alignments or matches, and increasing the word size gives less number of hits [33].

The implementation of proposed algorithm is done in Java. In this chapter, we will create code for proposed sequence alignment algorithm which we will run to do alignment of two sequences. The steps for creating program are (Fig. 10):

1. Define classes,
2. Define methods,
3. Write the Java code, and
4. Run and analyze the output.

The program involves data input from users, processing of that input, and producing the results. The input is taken from files containing the DNA sequences. Sequences are extracted from the files and then processed according to the program; that is, the query sequence (to be matches) is broken into segments. Size of segment is decided with choosing the word size. Word size is the size of segment in which the query sequence has to be broken. Then, these segments are matched with another sequence (with which to be matched) and alignment is done to find suitable alignments. Then, these alignments are extended to get higher scores. And then, the number of alignments which get higher score will be number of hits that are the output of this program.

5.4.1 Inputs

Two sequences are taken as inputs. These sequences are collected from NCBI database of nucleotides. Source of query sequence is Homo sapiens, and the sequence is as below:

ACAAGATGCCATTGTCCCCCGGCCTCCTGCTGCTGCTGCTCTCCGGG
GCCACGGCCACCGCTGCCCTGCCCCTGGAGGGTGGCCCCACCGGCCGA
GACAGCGAGCATATGCAGGAAGCGGCAGGAATAAGGAAAAGCAGCCT
CCTGACTTTCCTCGCTTGGTGGTTTGAGTGGACCTCCCAGGCCAGTGCC
GGGCCCCTCATAGGAGAGGAAGCTCGGGAGGTGGCCAGGCGGCAGG
AAGGCGCACCCCCCCAGCAATCCGCGCGCCGGGACAGAATGCCCTGCA
GGAACTTCTTCTGGAAGACCTTCTCCTCCTGCAAATAAAACCTCACCCA
TGAATGCTCACGCAAGTTTAATTACAGACCTGAA [31].

Source of other sequence is Tasmanian Mountain Shrimp, and the sequence is as below:

TCTTTAGATTTTATTTTTGGAGCTTGGTCTGGCATAGTAGGCACCGCC
CTAAGACTTATTATTCGGGCTGAATTAGGACAACCTGGTAGACTTATT
GTGATGATCAAATTTACAACGTGGTCGTAACAGCTCATGCTTTTGTGAT
ATTTTTTTTATAGTTATGCCCATTATAATTGGTGGATTTGGAAATTGACT
TTTCCCTTAATATTAGGTGCTCCTGATATAGCTTTTCCTCGTATAAATAA
TATAAGATTTTGACTTCTTCCACCTTCTTTAACTCTTCTCCTATCCAGA
GGAATAGTTGAAAGAGGTGTTGGCACAGGATGAACTGTTTATCCTCCTT
TAGCTGCTGGAATCGCCCATGC [31].

5.4.2 Java Packages

The packages that are used in the code are java.io and java.util. Java.io package means java input/output package which is used for taking input and producing output after processing input. All the operations related to input and output are controlled by some classes. And these classes are contained in this package. Java. util package has been used for scanner class (for taking input from user), hash table class (to create a hash table and storing segments in it), and enumeration (for retrieving the content of hash table).

5.4.3 Java Classes

The main class used for this program is ProposedAlgo which consists of all the methods defined in the program.

5.4.4 Java Methods

The program is developed in form of methods.

BuildMatrix()

This method creates a scoring matrix.

$$\text{double diagScore} = \text{score}[i-1][j-1] + \text{similarity}(i,j);$$
$$\text{double upScore} = \text{score}[i][j-1] + \text{similarity}(0,j);$$
$$\text{double leftScore} = \text{score}[i-1][j] + \text{similarity}(i,0);$$

First of all, the first row and column of matrix is filled with zero. Then for rest of cells in matrix, there are three possibilities of score. Suppose a cell x, for which score value can come from three directions (from upper cell, from left cell, or from right cell) and if the value comes from the diagonal cell, then score of cell x will be added to match or mismatch value. If the value comes from upper cell and left cell, then score of cell x is added to gap value.

$$\text{score}[i][j] = \text{Math.max}(\text{diagScore}, \text{Math.max}(\text{upScore}, \text{Math.max}(\text{leftScore}, 0)));$$

Then, find the maximum score value from scoring matrix and check that from which direction the maximum score value of cell x is obtained suing following code.

$$\text{prev\_cells}[i][j]| = \text{dr\_diag};$$

It means the diagonal direction gives maximum score to cell x.

$$\text{prev\_cells}[i][j]| = \text{dr\_left};$$

It means the left direction gives maximum score to cell x.

$$\text{prev\_cells}[i][j]| = \text{dr\_up};$$

It means the upper direction gives maximum score to cell x.

getMaxScore()

This method gets the maximum score from all the scoring matrixes by comparing the all cell values and storing the higher value to MaxScore. After comparing all cell values, the value stored in scoring matrix will be maximum score value in the scoring matrix. Then, this maximum score value which is in integer is normalized and converted to double value.

printAlignments()

This method prints the alignments having maximum score.

If the score comes from left direction, then align the two sequences using gaps as follows:

   p = printAlignments(i − 1, j, qstr1.charAt(i − 1) + aligned1, "_" + aligned2);

If the score comes from upper direction, then align the two sequences using gaps as follows:

   p = printAlignments(i, j − 1, "_" + aligned1, str2.charAt(j − 1) + aligned2);

If the score comes from diagonal direction, then align the two sequences using characters only and without gaps as follows:

p = printAlignments(i − 1, j − 1, qstr1.charAt(i − 1) + aligned1, str2.charAt(j − 1) + aligned2);

printAlignments(String)

This method extends the alignments obtained with printAlignments() method. These alignments are extended to get higher scores and then again aligned with same procedure before and final alignments we get with the higher scores will be the final suitable alignments. The number of extended alignments we get is the number of hits which we can see in the output at the end as number of matches = 25.

6   Experimental Evaluation

In Fig. 4.5, number of matches varies with varying word size. Threshold value is taken to be 10. With word size 7, number of matches are 16, and when word size = 11 and 15, then number of matches are zero (Figs. 11 and 12).

In Fig. 4.6, number of matches varies with varying word size. Threshold value is taken to be 3. With word size 3, number of matches are 120; when word size is 4, number of matches are 92; and when word size is 5, then number of matches are 74. That means when the word size is increasing, the number of matches has a score greater than the threshold decreases (Fig. 13).

In Fig. 4.6, number of matches varies with varying word size. Threshold value is taken to be 4. With word size 3, number of matches are 0; when word size is 4, number of matches are 58; and when word size is 5, then number of matches are 67.

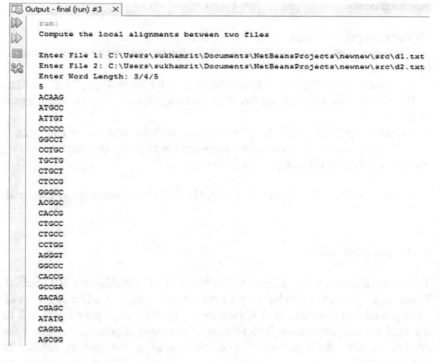

Fig. 11 Segments of query sequence

That means when the word size is increasing, the number of matches has a score greater than the threshold increases because of the threshold (Figs. 14 and 15).

In Fig. 4.6, number of matches varies with varying word size. Threshold value is taken to be 5. With word size 3, number of matches are 0; when word size is 4, number of matches are 0; and when word size is 5, then number of matches are 25. That means when the word size is increasing, the number of matches has a score greater than the threshold increases because of the threshold (Figs. 16 and 17).

7 Conclusion

DNA barcoding is a system for fast and accurate species identification which will make ecological system more accessible. It has many applications in various fields such as preserving natural resources, protecting endangered species. For species identification, sequence alignment is done in somewhat similar sequences.

Fig. 12 Extended alignments

Similarity search methods [1] have been used that include some already existing algorithms such as Smith–Waterman, BLAST, and FASTA. BLAST is being used for fast species identification but does not give accurate results like Smith–Waterman which is a very slow process. BLAST has been performed using a sequence to study the effect of word size on accuracy, and the results show that larger the word size, less will be number of hits and vice versa. So the idea is to combine the features of accuracy of Smith–Waterman and speed of BLAST algorithm. An algorithm is proposed and implemented on this idea with word size 3, 4, and 5, having improved accuracy with more number of matches. The number of matches varies with change in word size and the threshold value (Table 3).

```
Output - final (run) #4    X

    The alignments with the maximum score are:

    TTACA
    TTACA

    <<<<<<<<<<<<<<<<<<<<<<<<<<<<<<<<<<<<<<<<<<<<<<<<<<<<<<
    The extended alignmnets are:
    >>>>>>>>>>>>>>>>>>>>>>>>>>>>>>>>>>>>>>>>>>>>>>>>>>>>>>


    The maximum alignment score is: 4.0

    The alignments with the maximum score are:

    ACCT
    ACCT

    ACCT
    ACCT

    <<<<<<<<<<<<<<<<<<<<<<<<<<<<<<<<<<<<<<<<<<<<<<<<<<<<<<
    The extended alignmnets are:
    >>>>>>>>>>>>>>>>>>>>>>>>>>>>>>>>>>>>>>>>>>>>>>>>>>>>>>


    The maximum alignment score is: 0.0

    The alignments with the maximum score are:

    <<<<<<<<<<<<<<<<<<<<<<<<<<<<<<<<<<<<<<<<<<<<<<<<<<<<<<
    The extended alignmnets are:
    >>>>>>>>>>>>>>>>>>>>>>>>>>>>>>>>>>>>>>>>>>>>>>>>>>>>>>

Number of Matches:74
BUILD SUCCESSFUL (total time: 11 seconds)
```

Fig. 13 Number of matches

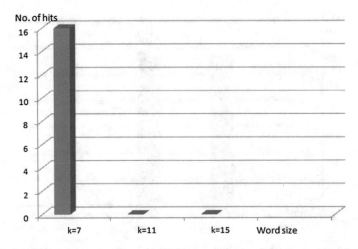

Fig. 14 Number of matches obtained with BLAST with varying word size and threshold = 10

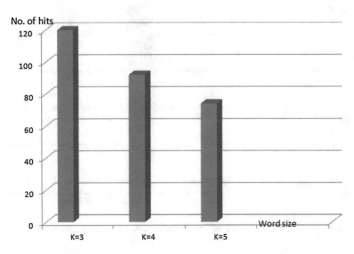

Fig. 15 Number of matches obtained with proposed code with varying word size and threshold = 3

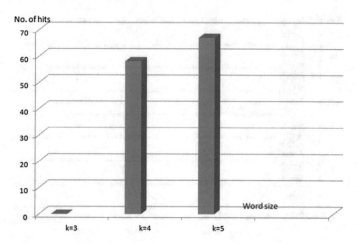

Fig. 16 Number of matches obtained with proposed code with varying word size and threshold = 4

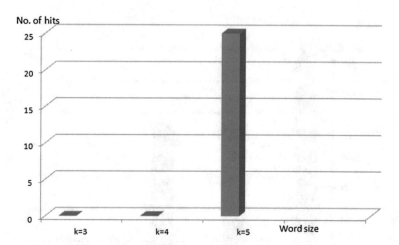

Fig. 17 Number of matches obtained with proposed code with varying word size and threshold = 5

Table 3 Different number of hits with different word size

Word size	Number of hits		
BLAST with threshold T = 10			
7	**16**		
11	**0**		
15	**0**		
Proposed code			
Threshold (*T*) →	**3**	**4**	**5**
3	**120**	**0**	**0**
4	**92**	**58**	**0**
5	**74**	**67**	**25**

References

1. V. Vaidya et al., A review of bioinformatics applications in breast cancer research. J.Adv. Bioinform. Appl. Res. **1**(1), 59–68 (2010)
2. N.M. Luscombe, What is bioinformatics? A proposed definition and overview of the field. J. Methods. Inf. Med. **40**(4), 346–358 (2001)
3. J. Cohen, Bioinformatics—an introduction for computer scientists. J.Comput. Biol. **36**(2), 122–158
4. M.B. Priyadarshni, Applications of bioinformatics. Available at: www.biotecharticles.com (2014)
5. Using DNA barcode to identify and classify living things (2014)
6. K. Sasikumar, C. Anuradha, DNA barcoding as a tool for algal species identification and diversity studies, **7**, 75–76 (2012)
7. P. Hebert, T.R. Gregory, The promise of DNA barcoding for taxonomy, **54**(5), 825–859 (2005)
8. F.C. Pereyra, M.G. Meckan, V. Wei, O. O'Shea, C.J.A. Bradshaw, C.M. Austin, Identification of Rays through DNA Barcoding: An application for ecologists, **7**(6), 1–10 (2012)
9. Identifying species with DNA barcoding. Available: www.barcodeoflife.org (2010)
10. R.D. Collins, R. Hanner, P.D.N. Hebert, The campaign to DNA barcode all fishes, FISH-BOL. J. Fish Biol. **74**, 329–356 (2009)
11. C. Ebach, C. Holdrege, DNA barcoding is no substitute for taxonomy, **434**, 697 (2005)
12. M.C. Ebach, C. Holdrege, More taxonomy, not DNA barcoding, **55**(10), 822–823 (2005)
13. M. Maheshwari, Sequence alignment, in *Introduction to Bioninformatics*, 1st edn. (New Delhi, 2008), pp. 164–196
14. Available at: www.dnabarcodes.org
15. W.J. Kress, D.L. Erickson, Introduction, in *DNA Barcodes Methods and Protocols* (Washington, DC, USA, 2012), pp. 3–10
16. P.M. Hollingsworth, Anti intellectualism in DNA barcoding enterprise. Zoologia. **3**(2), 44–47 (2007)
17. K.K. Dasmahapatra, J. Mallet, DNA barcoding: recent successes and future prospects. Hered. **97**, 254–255 (2006)
18. R.A. Collins, R.H. Cruik Shank, The seven deadly sins of DNA barcoding. Mol. Ecol. Res. 1–7 (2012)
19. C. Moritz, C. Cicero, DNA barcoding: promise and pitfalls. PLos Biol. **2**(10), 1529–1531 (2004)
20. P. Hebert, A. Cywinska, S.L. Ball, J.R. deWaard, Biological identifications through DNA barcodes. Proc. Biol. Sci. **270**(1512), 313–321 (2003)
21. C. Notredame, Recent evolutions of multiple sequence alignment algorithms. PLoS. Comput. Biol. **3**(8), 1405–1408 (2007)
22. D.P. Little et al., A comparison of algorithms for the identification of specimens using DNA barcodes: examples from gymnosperms. PLoS ONE. 1–21 (2007)
23. J. Waugh, DNA barcoding in animal species: progress, potential and pitfalls. Bio Essays. **29**(2), 188–197 (2007)
24. D.L. Dalton et al., DNA barcoding as a tool for species identification in three forensic wildlife cases in South Africa. Forensic. Sci. Int. **207**(1–3), 51–54 (2011)
25. Z.T. Nagy, T. Backeljau, M.D. Meyer, K. Jordaens, (2013), DNA barcoding: a practical tool for fundamental and applied biodiversity research. Zookeys. **5**(24)
26. C.B. Abilash, K. Rohitaktha, A comparative study on global and local alignment algorithm methods. IJETAE. **4**(1), 34–43 (2014)
27. A.G. Khallaf et al., DNA barcoding reveals a high level of mislabelling in Egyptian fish fillets. Food control. **46**, 441–445 (2014)

28. M.S. Rosenberg, Sequence alignment, in *Sequence Alignment Methods, Models, Concepts And Strategies*, 1st edn. (California, 2009)
29. D.E. Krane, M.L. Raymer, Data searches and pairwise alignments, in *Fundamental Concepts of Bioinformatics*, 1st edn. (New Delhi, 2006)
30. B. Bergeron, Pattern matching, in *Bioinformatics Computing* (New Delhi, 2003), pp. 302–339
31. Nucleotide Database, NCBI, NLM, NIH, 8600 Rockville Pike, Bethesda, MD 20894
32. EMBL-EBI, Pair wise sequence alignment, Wellcome Trust Genome Campus, Hinxton, Cambridgeshire, CB10 1SD, UK
33. By blast-help group, NCBI User Service, BLAST program selection guide. NCBI, NLM, NIH, 8600 Rockville Pike, Bethesda, MD 20894
34. NCBI News, GenBank surpasses one trillion total bases of publicly available sequence of data. NCBI, NLM, NIH, 8600 Rockville Pike, Bethesda, MD 20894 (2015)
35. J. Hanken, Trends Ecol. Evol. **18**(2) (2003)
36. Available at: http://seqcore.brcf.med.umich.edu/
37. H.A.I. Ramadan, N.A. Baeshen, Biological identification through DNA barcodes (2012)

Implementation of Ant Colony Optimization in Routing Protocol for Internet of Things

Prateek Thapar and Usha Batra

Abstract Internet of Things (IoT) is the next big step in evolution of Internet. IoT is a gradual upgrade of wireless sensor networks of embedded devices. Embedded sensor nodes can be directly addressed (IPv6) through the Internet owing to the 6LoWPAN adaptation layer. IETF working group Routing Over Low Power and Lossy Networks (ROLL) in 2012 introduced RPL–IPv6 Routing Protocol for low-power and lossy networks to provide efficient routing solution for 6LoWPAN abstraction layer running over IEEE 802.15.4 providing PHY and MAC for Low-Rate Wireless Personal Area Network. All the protocols including RPL insure minimal energy utilization of the sensors. RPL builds destination oriented directed acyclic graph (DODAG) using the objective function (OF). This objective function is responsible for fixation of rank of node and selection of best DAG and best parent. Large network of sensor nodes can be seen as a colony of ants. The Ant Colony Optimization is an important Swarm Intelligence technique under the paradigm of Computational Intelligence. It is inspired by collective intelligence of large number of homogeneous agents (ants). This optimization can be implemented toward selection of most efficient route and hence the preparation of DODAG. In this paper, we implement Ant Colony Optimization in RPL and present the results based on simulation using COOJA simulator in Contiki operating system.

1 Introduction

The research on embedded systems and wireless sensor networks along with the Internet Protocol version 6 (IPv6) has resulted in the scenario wherein the wireless sensors can have independent IP addresses and can be accessed individually through Internet. This revolution has brought mechanical devices and electronic

P. Thapar (✉) · U. Batra
GD Goenka University, Gurgaon, India
e-mail: prateek.thapar@gdgoenka.ac.in

U. Batra
e-mail: usha.batra@gdgoenka.ac.in

© Springer Nature Singapore Pte Ltd. 2018
B. Panda et al. (eds.), *Innovations in Computational Intelligence*, Studies in Computational Intelligence 713, https://doi.org/10.1007/978-981-10-4555-4_10

sensors together on the Internet and is called as Internet of things (IoT). The 'Things' in IoT are power-constrained devices running on low power and over lossy networks. This peculiarity of sensor nodes has called for a separate protocol stack for its implementation. IEEE 802.15.4 provides Physical and Mac layer for Low-Rate Wireless Personal Area Network [1]. IPv6 over low-power wireless personal area networks—6LoWPAN [2] abstraction layer condenses the IPv6 address for communication, and RPL protocol is used for routing of messages. All the protocols are designed with an aim to conserve energy so that the life of constrained nodes can be increased.

RPL handles the packet processing and forwarding separately from the route optimization. RPL uses direction oriented directed acyclic graph (DODAG) for creation of network topology. Each node communicates with more than one node but has a single preferred parent for the formation of DODAG. The selection of preferred parent is based on an objective function defined in RPL. The RPL implementation in Contiki operating system provisions objective functions with metric based on expected transmission count (ETX) and hop count.

Computational Intelligence is a subfield of Artificial Intelligence, dealing with the sub-symbolic techniques with focus on strategy and outcome of combined intelligence of a group of lesser intelligent agents. Ant Colony Optimization (ACO) is one of the prominent techniques of Computational Intelligence used in research. Ant Colony Optimization has been used for solving traveling salesman problem and for cost optimization.

In this paper, we present an approach whereby ACO is used to optimize the path energy consumption, i.e., the sum of energy consumed by transmitting node during transmission and energy consumed by listening node during listen process, over the ETX metric-based objective function. The experiments are conducted in COOJA simulator on Contiki OS, and the results of ETX-based objective function are compared with the results obtained after implementation of ACO.

The remaining part of the paper is organized as follows. Section 2 covers the detailed functioning of the RPL. The Ant Colony Optimization is explained in Sect. 3. The motivation and related research is presented in Sect. 4. We present the proposed methodology in Sect. 5 and preliminary results in Sect. 6. Section 7 concludes the work with the emphasis on future work.

2 Routing Protocol for IoT

IPv6 Routing Protocol for low-power and lossy networks (RPL) [3] is a distance vector routing protocol which operates as the Network Layer protocol in the protocol stack for communication among Internet of Things. IETF working group (WG) Routing Over Low Power and Lossy Networks (ROLL) in 2012 introduced RPL to provide efficient routing solution for 6LoWPAN running over IEEE 802.15.4 PHY and MAC layer. RPL uses destination oriented directed acyclic graph (DODAG) as the topology for communication. There can be more than one

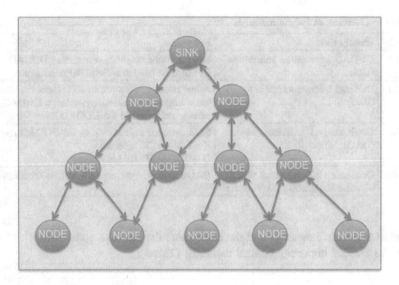

Fig. 1 Representation of a DODAG

parents to a node toward the sink/server called DAG root in this context. Optimal routes are constructed using preferred parent between sink and other nodes for both upward and downward traffic. Redundant equivalent routes are also kept for reliability. Within one RPL Instance, multiple DODAGs can be present each with same of different objective functions. These DODAGs are constructed and maintained through DODAG information object (DIO) messages (Fig. 1).

2.1 RPL Traffic Flow

RPL supports MultiPoint-to-Point (MP2P), Point-to-MultiPoint (P2MP), and Point-to-Point (P2P) as the traffic flow techniques. MP2P traffic flow techniques mostly belong to the instances wherein a single node provides a sink for many child nodes. Such sink node attracts the upward traffic. Whereas, the P2MP traffic techniques employs the downward traffic from sink to multiple child nodes such as firmware upgrades. P2P packets flow upwards through preferred parent till the node is attained under which the destination node exists.

2.2 Control Messages

RPL uses Internet Control Message Protocol Version 6 (ICMPv6) [4] with TYPE = 155. RPL control message is identified with the code that specifies the

Table 1 Overview of control messages

Code	Identification	Function
0x00	DODAG information solicitation (DIS)	Used to probe neighbors for nearby DODAGs. Can also be used to solicit DIO messages
0x01	DODAG information object (DIO)	Allow nodes to discover an RPL Instance, learn the configuration parameters, select a DODAG parent, and maintain the DODAG
0x02	Destination advertisement object (DAO)	Send the node information to the DODAG root K—DAO-ACK expected D—DODAGID field is present
0x03	Destination advertisement object —acknowledgment (DAO-ACK)	DAO sequence—corresponds to DAO message

type of the control message and the base where the message goes. Following table denotes various important control messages (Table 1).

2.3 Objective Function

RPL separates the packet processing and forwarding from the route optimization. The routes in an RPL Instance are optimized using an objective function. An RPL implementation has one or more RPL Instances, while one RPL Instance will have multiple DODAGs but with a single objective function. The objective function decides the preferred parent, and the calculations for it are used to compute the rank of the node. The rank of a node determines the position of the node in a DODAG. RPL in Contiki [5] operating system supports two types of objective functions, viz, ETX-based objective function and the OF0 objective function which is based on the hop count with respect to the root node.

2.4 Expected Transmission Count (ETX)

ETX [6] refers to the estimated number of transmissions from node X to node Y as the number of times a frame needs to be transmitted over a link before being acknowledged. ETX is calculated as the ratio between number of acknowledgements received and the number of packets transmitted. The default objective function used in RPL minimizes the ETX. The parent with least ETX is selected as the preferred parent in the DODAG. Contiki OS computes ETX as the count of number of retransmissions without considering the number of packets being dropped by MAC layer. Hence, the basic component affecting ETX emerges as the quality of the link which is implemented as an exponentially weighted moving average (EWMA) filter [7]. Authors in [8] show that in default RPL implementation (ETX based) the power consumption is nearly doubled if sink is moving than the power consumption if the sink is placed at a fixed location.

2.5 Trickle Algorithm

The function of trickle algorithm [9] is to dynamically adjust the transmission window for nodes to share their status information based on the changes in the current setup or link status. Trickle algorithm is primarily important for communication between lossy and low-power nodes, hence is employed in RPL. Each node in default mode of RPL broadcasts its rank based on a timer governed by the trickle algorithm. The timer increases exponentially till the time no inconsistency in the network is detected. On detection of a link failure or other inconsistency, the timer is reset and the nodes broadcast their ranks through DIO messages.

3 Ant Colony Optimization

Computational Intelligence is the field that describes the sub-symbolic techniques with a focus on strategy and outcome. Computational Intelligence broadly covers subfields such as Swarm Intelligence, Fuzzy Systems, Artificial Neural Networks, and Evolutionary Computation. Swarm Intelligence is the field in which the focus is on collective intelligence by a large group of lesser intelligent agents. The collective intelligence of these agents such as ants, birds, bacteria, fish, and bees is attained through their interaction and cooperation. These agents interact naturally to fulfill their basic necessity of food and relocation through use of pheromones in ants, dancing in bees and chemical gradients in bacteria. Swarm Intelligence has majorly being studied through Ant Colony Optimization, Particle Swarm Optimization, Bees Algorithms, and Bacteria Foraging Optimization.

Ant Colony Optimization [10] is the study of probabilistic algorithms based on foraging behavior of ants and implementation of these algorithms in search and optimization domain. Ants lay down pheromone on the path they travel from colony to the food source. Ants on the forward as well as backward journey deposit the pheromone. The pheromone decays in the environment. At every junction, there is a probability of selection of one of the paths. In the event, when all the probable routes are new and no pheromone is deposited on any of the routes, the probability of selection of all the routes is equal. As the ants start traveling forward and backward, depending on the frequency of travel, the pheromone deposits on the shorter route increases. This increase in pheromone increases the probability of the selection of a particular route for selection at the edge. The pheromone deposited on the longer routes decays over a period of time, and eventually one shortest route emerges. This mechanism of pheromone communication between ants for selection of a good path is called stigmergy.

4 Motivation and Related Research

RPL implementation in Contiki OS has ETX-based minimum rank with hysteresis objective function (MRHOF) [11] with an option to change to energy based on node energy. Objective function zero (Of0) is another objective function implemented in Contiki that is purely based on rank of the neighboring nodes toward selection of parent/preferred parent. Researchers worldwide have conducted several studies toward evaluation of performance of RPL using different simulation environments, frameworks, and scenarios. Authors in [12] evaluate the performance of ETX and Of0 objective function. Their results show that ETX objective function performs better than Of0 in terms of power consumption and average path cost. Author in [13] shows that ETX (3.7%) is using percentage RADIO ON time (deducible to energy consumption) more efficiently than Of0 (4.7%) during an hour simulation.

Further the researchers have proposed various models to improve the performance of RPL under different scenarios. Authors in [14] propose received signal strength indicator (RSSI)-based routing metric instead of ETX for lossy networks. Their experiments show improvement in transmission performance in 6LoWPAN. Similar agenda of improving the performance of RPL yet different approaches has been published as [15] wherein the authors propose Scenario-based Heuristic for Robust Shortest Path tree that is experimentally proved to be resilient to link channel (ETX) variability.

A limited research has been conducted on swarm intelligence-based routing protocols for Internet of Things (IoT). Authors in [16] have proposed an objective function based on Ant Colony Optimization and have compared with the ETX-based objective function. Their QoS_RPL objective function aims to use nodes with higher residual energy so that the residual power of all nodes in network reduces proportionately. This increases the network lifetime by approximately 10%. However, this proposal shows reduced transmission performance with respect to the good old ETX-based objective function.

Current research appears to validate the view that there is definitely scope for improvement in present implementation of objective functions in RPL. A closer look at the literature reveals ETX-based objective function performs better than OF0 in terms of energy consumptions and transmission rate in spite of the fact that the link quality is highly variable in real scenario for IoT. Although there have been little research in swarm intelligence implementation in RPL, yet there is a scope of this field of research for routing in IoT.

5 Proposed Methodology

In this paper, we propose an adaptation of Ant Colony Optimization for providing a correction to the ETX metric-based MRHOF. The ACO is proposed to optimize the path energy that is calculated as the sum of transmission energy consumed by transmitting node and the energy consumed by listening node in the listen process. An additional dedicated node is placed in the network that gathers the power consumption data, determines the lowest path cost using ACO, and calls an additional function in the rpl-dag.c designed to update the preferred parent of the communicating nodes. Contiki OS and its simulator COOJA have been used for experimentation of the proposed methodology.

5.1 RPL in Contiki OS

RPL has been implemented in Contiki OS by Swedish Institute of Computer Science through following files in /core/net/rpl:

(a)	rpl.c	RPL main implementation file as per [3]
(b)	rpl.h	Public API declarations for ContikiRPL
(c)	rpl-conf.h	Public configuration and API declarations
(d)	rpl-dag.c	Logic or directed acyclic graph
(e)	rpl-ext-header.c	Management of extension headers for ConitkiRPL
(f)	rpl-icmp6.c	ICMPv6 I/O for RPL control messages
(g)	rpl-mrhof.c	Implementation of the minimum rank with hysteresis objective function [11]
(h)	rpl-of0.c	Implementation of RPL objective function zero [17]
(i)	rpl-private.h	Private declaration for ContikiRPL
(j)	rpl-timers.c	RPL Timer Management

5.2 Powertrace

Powertrace [18] is an application developed by Swedish Institute of Computer science in Contiki OS for network-level power profiling. It is used to empirically evaluate and compare the energy consumption. This application uses *Energest.c* and *Energest.h* files from /core/sys for calculation of energy consumption. The results obtained vide use of powertrace include energy consumed by cpu, cpu in low power mode (lpm), during transmission and during listen. Powertrace is based on linear power model through which the power at each active power state is estimated and added to attain the instantaneous power. The instantaneous power consumption and energy consumption are described as:

$$P_{\text{system}}(t) = \sum_{m,n} P_{m,n} S_{m,n}(t) \tag{1}$$

where

$P_{\text{system}}(t)$ Power consumption of system at instance 't'
$P_{m,n}$ Power consumed by 'm' component in 'n' state
$S_{m,n}(t)$ state i.e., 0 or 1

$$E_{\text{system}} = \sum_{m,n} P_{m,n} T_{m,n} \tag{2}$$

where

E_{system} Energy consumption of system
$P_{m,n}$ Power consumed by 'm' component in 'n' state
$T_{m,n}$ Time for which 'm' component has been in 'n' state

$$\text{Energy Consumption} = \frac{\text{Energest\_Value} \times \text{Current} \times \text{Voltage}}{\text{RTIMER\_SECOND} \times \text{Run Time}} \tag{3}$$

5.3 Ant Colony Optimization

Ant Colony Optimization (ACO) is a subfield of swarm intelligence and a technique based on stigmergy. Authors in [19] have explained the procedure, how Ant Colony Optimization can be used for minimization of a cost function. This procedure has been implemented on energy readings for selection of path with minimum consumption of path energy (energy consumed by edge nodes in transmission and listening of the packets, respectively). The implementation of Ant Colony Optimization is performed through ACO parameter value as depicted in Table 2 and functions mentioned hereunder:

Table 2 ACO parameter values

Parameter	Value
α—influence of pheromone in direction	3.0
β—influence of adjacent node distance	2.0
ρ—pheromone decrease factor	0.01
Q—pheromone increase factor	2.0

Probability of Pheromone Deposit

$$P^k_{xy} = \frac{\left(T^\alpha_{xy}\right)\eta^\beta_{xy}}{\sum_{x=\text{allowed}}\left(T^\alpha_{xz}\right)\left(\eta^\eta_{xz}\right)} \tag{4}$$

where

T^α_{xy} amount of pheromone deposited for transmission from state $x \rightarrow y$.

η^β_{xy} desirability of state transition xy typically $1/d_{xy}$ where d is the distance and $\beta > 1$.

T_{xz} and η_{xz} attractiveness and trail level for other possible state transition.

Pheromone Update

$$T_{xy} \leftarrow (1 - \rho)\, T_{xy} \sum_k \Delta T^k_{xy} \tag{5}$$

where

T^k_{xy} amount of pheromone deposited for state transition xy.

ρ Pheromone evaporation coefficient

ΔT^k_{xy} amount of pheromone deposited by kth ant.

5.4 Simulation Environment

For the evaluation of the proposed methodology, COOJA simulator in Contiki operating system is used. We use the Sky motes with one mote as UDP Server and rest of the motes as UDP Clients. The network diagram, consisting of total 20 nodes including Node 1 as the UDP Server and Node 2–20 as UDP Clients, is shown in Fig. 2. We have used the following tools while development:

(a)	Instant Contiki 3.0 that is available as image (.iso) file with preinstalled tools
(b)	VMware Workstation 12 Player for running the Instant Contiki 3.0 virtual machine
(c)	Wireshark as network protocol analyzer
(d)	Matplotlib, a python-based 2D graph plotting library
(e)	Sublime Text Build 3114 as text editor for writing code and customization of preexisting code files

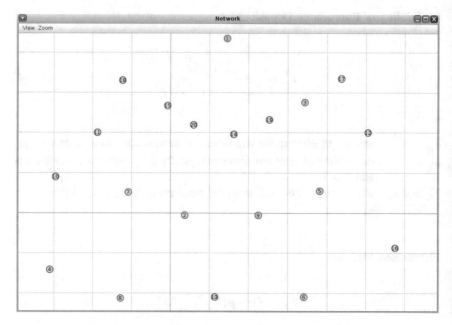

Fig. 2 Network diagram for simulation

Table 3 COOJA simulation settings

Settings	Value
Radio medium	DGRM (so that link can be degraded manually)
Mote type	Sky mote
Contiki process/firmware	Modified udp-server.c/udp-client.c
Radio messages	6LoWPAN analyzer with PCAP
Radio interface	IEEE 802.15.4

Directed graph radio medium (DGRM)-based simulation is performed in COOJA simulator with settings as shown in Table 3.

5.5 Workflow

The simulation of the network starts with Rime initialization. During the simulation, as the nodes wake up, they broadcast DIS messages. The neighboring nodes on receipt of DIS message send out DIO message. The nodes that receive the DIO message process it for choosing the preferred parent. The ETX-based MRHOF objective function is used for initial preferred parent selection. This process continues till the node decides to join a DAG and advertise itself upward in the DAG through DAO message to the parent. The parent then acknowledges the child node with DAO-ACK message. The child node thereafter broadcasts the DIO message containing his rank in the DAG. Along with the initialization of the network, powertrace has also been initialized in nodes so that the energy consumption can be monitored. These powertrace output readings are converted to link-based path energy consumption in form of a csv file. At predefined regular intervals, the energy consumption readings are converted from csv to a graph and processed by program based on Ant Colony Optimization. This program determines the path from each client to server expending minimum energy. The output from this program is used to update the preferred parent of each node. In process of updating this information, the preferred parent is only altered if the rank of the new parent is better or equal to the preexisting preferred parent. The energy consumption statistics are correlated for obtaining the results.

6 Preliminary Results

We evaluated the performance of ETX metric-based MRHOF with and without implementation of the proposed methodology. The simulations were carried out for a fixed period of time of 120 min each with ACO implementation and without it. The links were established using DGRM model whereby initial specification of links is 100% Rx Ratio, −10.0 RSSI, 105 LQI, and 0 ms Delay. However, during the course of simulation, the link degradations were simulated multiple times. The links were degraded with 50% Rx Ratio, −70.0 RSSI, 70 LQI, and 0 ms Delay as the link characteristics. These degradations were selectively carried out on DODAG links so that the objective function is called multiple times to establish a new DODAG.

During the simulation, it was observed that the RPL was unable to modify the DODAG that was established initially for some of the nodes for a long period of time. However, when the Ant Colony Optimization was carried out on the path energy and the preferred parent was updated, this change was accepted immediately and during the first pass, all the nodes updated their parent and new DODAG was established with better links. The details are shown in Table 4. In another set of

Table 4 No. of degraded links being used

Check No.	Simulation time (min)	Elapsed time since degradation	Objective function	No. of degraded links being used
1	20	10	Without ACO	03
			With ACO	Nil
2	40	20	Without ACO	03
			With ACO	01

Fig. 3 Energy consumption during transmit (without ACO)

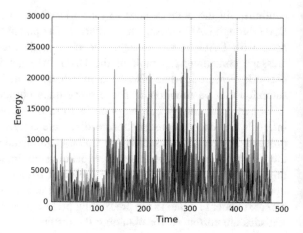

Fig. 4 Energy consumption during transmit (with ACO)

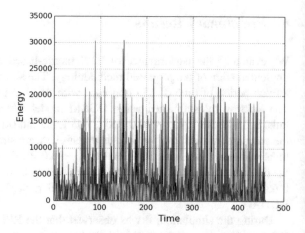

simulations, it was observed that with the passage of time, the number of nodes changing over to the better link reduced when plain ETX metric-based MRHOF was implemented. This degrades the quality of service given by RPL.

The energy consumption during transmission process for all the nodes for the simulation is shown in Fig. 3 for implementation without ACO and Fig. 4 for proposed methodology. The graph clearly indicates that initially the network is consuming more energy in transmission when ACO is applied to the network as more amount of control messages are sent over the network to change the links of DODAG from degraded ones to more efficient links. This optimization is successful in stabilizing the network transmissions over a period of time. Hence, the later half of the graph in Fig. 4 indicates fairly stable transmission energy consumption.

7 Conclusions and Future Work

The prime focus in IoT is to conserve energy of sensor nodes. In order to accomplish the same, the simulations show that RPL does not encourage the change of parent and hence the node continues to communicate through degraded links. This scenario works well in case, where the communication requirement is low and the link status remains fairly stable. However, the cases wherein the data transfer requirements are high and links quality keeps changing, plain ETX-based MRHOF is unable to give better quality of service. In this research paper, we have implemented ACO over ETX-based MRHOF and provided a mechanism to change over the ailing link to another efficient link. We plan to focus our next goal in this research toward designing of a new objective function and trickle algorithm based on energy consumption statistics of various nodes in the network.

References

1. Low-rate wireless personal area networks (LR-WPANs). IEEE 802.15.4, Sept 2011
2. N. Kushalnagar et al., IPv6 over low-power wireless personal area networks (6LoWPANs). IETF RFC 4919, Aug 2007
3. T. Winter, A.B.P. Thubert, T. Clausen et al., RPL: IPv6 routing protocol for low power and lossy networks. IETF RFC 6550, Mar 2012
4. A. Conta et al., Internet control message protocol (ICMPv6) for the internet protocol version 6 (IPv6) specification. IETF RFC 4443, Mar 2006
5. A. Dunkels, B. Gronvall, T. Voigt, Contiki-a lightweight and flexible operating system for tiny networked sensors, in *29th Annual IEEE International Conference on Local Computer Networks, 2004.* IEEE (2004)
6. J.P. Vasseur et al., Routing metrics used for path calculation in low-power and lossy networks. IETF RFC 6551, Mar 2012
7. S. Dawans et al., On link estimation in dense RPL deployments, in *Proceedings of the International Workshop on Practical Issues in Building Sensor Network Applications* (IEEE SenseApp, 2012), Florida, USA, Oct 2012
8. I. Wadhaj et al., Performance evaluation of the RPL protocol in fixed and mobile sink low-power and lossy-networks, in *2015 IEEE International Conference on Computer and Information Technology; Ubiquitous Computing and Communications; Dependable, Autonomic and Secure Computing; Pervasive Intelligence and Computing (CIT/IUCC/DASC/PICOM).* IEEE (2015)
9. P. Levis et al., The trickle algorithm. IETF RFC 6202, Mar 2011
10. M. Dorigo, M. Birattari, T. Stutzle, Ant colony optimization. IEEE Comput. Intell. Mag. **1**(4), 28–39 (2006)
11. O. Gnawali, P. Levis, The minimum rank with hysteresis objective function. IETF RFC 6719, Sept 2012
12. R. Sharma, T. Jayavignesh, Quantitative analysis and evaluation of RPL with various objective functions for 6LoWPAN. Indian J. Sci. Technol. **8**(19), 1 (2015)
13. H. Ali, A performance evaluation of RPL in Contiki (2012)
14. T.-H. Lee, X.-S. Xie, L.-H. Chang, RSSI-based IPv6 routing metrics for RPL in low-power and lossy networks, in *IEEE International Conference on Systems, Man, and Cybernetics (SMC).* IEEE (2014)

15. I.A. Carvalho et al., A scenario based heuristic for the robust shortest path tree problem. *IFAC-PapersOnLine* **49**(12): 443–448 (2016)
16. B. Mohamed, F. Mohamed, QoS routing RPL for low power and lossy networks. Int. J. Distrib. Sens. Netw. **2015**, 6 (2015)
17. P. Thubert, Objective function zero for the routing protocol for low-power and lossy networks (RPL). IETF RFC 6552, Mar 2012
18. A. Dunkels et al., Powertrace: network-level power profiling for low-power wireless networks (2011)
19. J. Brownlee, Clever algorithms. Nature-inspired programming receipes (Chap. 6), pp. 237–260 (2011)

A Lexical and Machine Learning-Based Hybrid System for Sentiment Analysis

Deebha Mumtaz and Bindiya Ahuja

Abstract Micro-blogs, blogs, review sites, social networking site provide a lot of review data, which forms the base for opinion mining or sentiment analysis. Opinion mining is a branch of natural language processing for extracting opinions or sentiments of users on a particular subject, object, or product from data available online Bo and Lee (Found Trends Inf Retrieval 2(1–2):1–135, 2008, [1]). This paper combines lexical-based and machine learning-based approaches. The hybrid architecture has higher accuracy than the pure lexical method and provides more structure and increased redundancy than machine learning approach.

Keywords Lexical analysis · Machine learning · Sentiment · Hybrid approach

1 Introduction

A massive quantity of data is generated daily by social networks, blogs, and other media on the World Wide Web. This enormous data contains vital opinion-related information that can be exploited to profit businesses and other scientific and commercial industries. However, manual mining and extraction of this information are not possible, and thus, sentiment analysis is required. Sentiment analysis is an application of natural language processing and text analytics for extracting sentiments or opinions from reviews expressed by users on a particular subject, area, or product online [1]. Its main aim is to classify each sentence or document as positive, negative, or neutral. The techniques employed so far can be categorized into lexical-based approach and machine learning approach.

The lexical approach is based on the supposition that the combined polarity of sentences is the summation of individual word polarity [2]. This approach classically employs a lexical or dictionary of words which are pre-labeled according to their polarity. Then, every word of the given document is compared with the words

D. Mumtaz (✉) · B. Ahuja
MRIU, Faridabad, Haryana, India
e-mail: deebhamumtaz@gmail.com

© Springer Nature Singapore Pte Ltd. 2018
B. Panda et al. (eds.), *Innovations in Computational Intelligence*, Studies in
Computational Intelligence 713, https://doi.org/10.1007/978-981-10-4555-4_11

in the dictionary [3]. In case the word is found, then its polarity strength value is added to the total polarity score of the text, and hence, the orientation is finally obtained. The main benefit of the lexical approach is that it is a simplistic method, fast, and works well for precise, small dataset. However, its disadvantages are low accuracy, low precision, low recall, and low performance in complex datasets.

In machine learning technique, a group of feature vector is selected and a labeled dataset, i.e., training data, is made available for training the classifier. The classifier can then be used to classify the untagged corpus (i.e., test data set). The choice of features is critical for the performance of the classifier [4]. Usually, a range of unigrams (single words) or n-grams (two or more words) is selected as feature vectors. Apart from these, the features may comprise of the frequency of opinion words, the frequency of negation words, strength of words, and the length of the text. Support vector machines (SVMs), max entropy, and the Naive Bayes algorithm are the most important classification algorithms used [5].

A. Naive Bayes classifier is based on Bayes' theorem, which assumes that the value of a particular feature is not dependent on the value of any other feature, i.e., each feature is independent. This model is easy to build, simple, and helpful for very large datasets, with no complicated iterative parameter estimation present [6, 7].

B. Maximum entropy is a technique for obtaining probability distributions of given data. The basic principle is that when no information is known, then the distribution should have maximal entropy [6]. Labeled training data provides constraints on the distribution, determining where to have the minimal nonuniformity [7].

C. Semi-supervised learning is a class between unsupervised learning and supervised learning that makes use of unlabeled data. Many researchers have found that unlabeled data, when used in combination with a small quantity of labeled data, can produce a significant enhancement of learning accuracy [7, 8].

D. Support vector machines are supervised techniques coupled with learning algorithms that examine data used for classification. Given a set of training examples, each distinctly labeled for belonging to one of the types. An SVM training algorithm builds a model that allot novel examples into one category or the other, making it a binary linear classifier which is non-probabilistic [3].

The advantage of machine learning approach is its high accuracy and high precision, and it works well on complex datasets. But the disadvantage is that the final output quality depends on the quality of training dataset, and it is less structured, sensitive to writing style, and has low performance in cross-style setting [9].

The proposed approach is based on combining the two techniques. Compared to lexical system, it achieves considerably high accuracy, and with respect to machine learning approach, it offers more structure, readability, and reliability. The remaining paper is organized as follows. In Sect. 2, a review of the related work is discussed. In Sect. 3, the proposed system based on hybrid approach is introduced. In Sect. 4, the experimental results are given and conclusion is presented in Sect. 5.

2 Related Work

Pang and Lee [1] assessed and compared the various supervised machine learning algorithms for classifying the opinions of movie reviews. They used learning algorithms such as Naive Bayes (NB), maximum entropy (ME), and support vector machine (SVM). With a straightforward algorithm using SVM, they trained on bag-of-words attributes and obtained 82.9% accuracy, which was later enhanced in their future work [10] to 87.2%. However, other researchers have found [11] that such a simple design experiences challenges of domain, time, and style dependencies. Additionally, it gives just a general sentiment value for each sentence exclusive of any clarification as to what people exactly loved or hated and up to what extent they did so.

Hu and Liu [12] put forward a bi-step process for sentence-level opinion mining, which was then enhanced by Popescu and Etzioni [12]. According to this process, opinion mining could be divided into two stages: aspect categorization and measurement of sentiment strength for each feature. Feature selection is a vital step since the aspects set up the domain in which the sentiment was expressed. Hu and Liu [12] found that aspect identification can be carried on by choosing the most commonly used nouns and noun phrases. But reviews from customers are frequently short, casual, and not precise, which makes this job very difficult. In order to overcome this challenge, labeled sequential rules (LSR) can be used [13], where such a rule is fundamentally a unique kind of sequential pattern.

Usually, most sentiment analysis techniques are based on either pure lexical or learning methods. However, with growing recognition of numerous generative probabilistic models, based on latent Dirichlet allocation (LDA), some attempts have been made to integrate lexical method into machine learning system [14, 15]. Davidov et al. [16] researched on the presence of hashtags and emoticons in opinion mining. Agarwal et al. [17] studied novel features for sentiment analysis of Tweets. Mittal et al. [18] proposed a hybrid system combining Naïve Bayes and lexical method and acquired an accuracy of 73%. They also incorporated tools for spelling correction (e.g., gooood as good) and slang handling. Malandrakis [19] combined lexical and max entropy algorithm for sentiment analysis with POS tagging, emoticons, and n-grams. Harsh [20] combined SVM and Naive Bayes algorithm and acquired an accuracy of 80%.

3 Proposed Approach

The proposed system is a combination of both lexical and machine learning techniques, hence exploiting the best characteristics of both in one. For this purpose, we choose to utilize a support vector machine (SVM) and empower it with an English lexical dictionary AFINN [8]. Our hybrid system follows the conventional four steps, namely data collection, preprocessing, training the classifier, and classification.

Through the following sections, we shall discuss each step in detail, one at a time. Figure 1 shows the system architecture of the proposed approach.

A. Data collection:

The first step is the collection of opinion data; we can get the data from various online sources such as Twitter, Facebook, Pinterest, Rotten Tomatoes. In this system, we required data to be labeled as positive, negative, and neutral for training and result analysis. We obtained a part of the data from Bing [5], which has a collection of product reviews (camera, phone, etc.) acquired from Amazon.com.

B. Preprocessing:

The opinion sentences may contain non-sentimental data such as URLs, hashtags "#", annotation "@", numbers, stop words. Preprocessing is the vital step to remove the noisy data.

The preprocessing module filters data and removes unnecessary characters, converts alphabets to lower case, and splits up sentences into individual words.

(1) Erratic casting: In order to tackle the problem of posts containing various casings for words (e.g., \BeAUtiFuL "), the system transforms all the input words to lower case, which provides consistency.

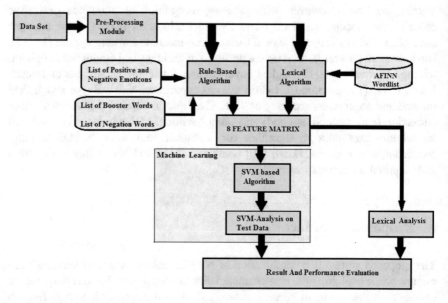

Fig. 1 Architecture of the proposed system

(2) Stop words: The presence of common words and stop words such as "a", "for", "of", "the" is usually ignored in information retrieval since their presence in a post does not provide any useful information in classifying a document.
(3) Language detection: Using NLTK's language detection feature, all sentences can be separated into English and non-English data.
(4) Split sentence: The given opinion sentence is split into separate individual words, and then, the lexical analysis is performed.

C. Rule-based Algorithm

In the rule-based algorithm, a list of positive and negative emoticons, booster words, and negation words is provided. The training data is compared with these lists and match is done. The frequencies of each of these words are obtained and used as aspect for machine learning algorithm.

(1) Emoticons: It is a symbolic illustration of a facial expression made with letters, punctuation marks, and numbers generally included in text online [21]. Many people make use of emoticons to convey emotions, which makes them very helpful for sentiment analysis [4]. A collection of about 30 emoticons, including "=[", "=)", ":)", ":D", "=]", ":]", ":(", "=(" are categorized into positive and negative.
(2) Negation words: The data contains negation words such as "no", "not", "didn't", which have a huge effect on the polarity of the sentence [9]. These words when present with a sentiment word reverse the polarity of the sentiment. For example, "I didn't like the movie" has a negative orientation even though it contains positive sentiment word "like."
(3) Booster words: Words such as "very", "highly", "most" are known as booster words; these when combined with other sentiment words enhance or boost the expression. For example, "The girl is very beautiful" gives a high positive sentiment score.

D. Lexical Algorithm

This algorithm is based on the hypothesis that the polarity of the sentence equals the summation of the polarity of individual words [7]. It uses a dictionary of positive and negative opinion words based on which the polarity of each word is obtained. In this research work, AFINN [8] wordlist is used in order to analyze the sentence content. AFINN wordlist contains about 2475 words and phrases which are rated from very positive [+5] to very negative [−5]. The words which have the score between +3 and +5 are categorized as very positive, the score of +1 to +2 categorized as positive, the score of −1 to −2 grouped as negative, and the score of −3 to −5 grouped as very negative.

The frequencies of the opinion words in each of the four categories are calculated and stored as a matrix. Also, the positive and negative emoticon frequency obtained from the rule-based algorithm is taken. Then, the basic algorithm for lexical analysis is applied, wherein the resultant polarity of the sentence is equal to the difference between the sum of positive and negative opinion words and emoticons. If the value is positive, then the sentence has positive sentiment; if the score is negative, then the polarity is negative or else the sentence is neutral.

E. Machine Learning Module

In the previous step, two outputs are obtained: the lexical analysis of the dataset as positive, negative, or neutral and a 7-feature matrix. This matrix consists of the frequencies of very positive words, very negative words, positive words, negative words, positive emoticons, negative emoticons, negation words, and booster words. It acts as a base for feature selection in machine learning. Machine learning as already discussed trains the machine on the training dataset and then tests it on the testing dataset. In machine learning, labeled examples are employed to learn classification. The machine learns by using features obtained from these examples.

Support vector machines (SVMs) are supervised techniques combined with learning algorithms that inspect data intended for categorization. Support vector machines show high efficiency at conventional data classification, usually better than Naive Bayes [4]. In the bi-class case, the basic idea behind the training procedure is to find a maximum margin hyperplane, represented by vector w, that not only separates the document vectors in one class from those in the other, but for which the separation, or margin, is as large as possible.

F. Result Analysis

The last step is the analysis of results obtained by pure lexical and hybrid approach (Fig. 2), which can be performed by evaluating confusion matrix, accuracy, recall, and performance estimation.

4 Results

The results of both of the existing classifier and proposed hybrid classifier are presented and compared in the format of deployment. Generally, performance is estimated by employing three indexes: accuracy, precision, and recall. The fundamental method for calculating these relies on confusion matrix. A confusion matrix is a matrix utilized to evaluate the performance of a classification model on a set of test data for which the true values are known [7]. It can be represented as given in Table 1.

Fig. 2 Algorithm proposed

Algorithm Proposed
1) Input Labeled Review Sentences *R*
2) Perform Pre-processing to filter noisy data, remove numbers, stop words, non-English words. Convert to lower case.
3) For each *R*, separate each word *W, do*
   # *Lexical Analysis*
   1. Input list of positive and negative sentiment words
   2. Initialize *Score, PosWord, NegWord, VPos, VNeg,* variables to *Null*;
   3. Compare each word to list of sentiment words.
   4. Get the 4-feature matrixes containing frequency of very positive, positive, very negative and negative words in the sentences.
   5. Calculate *Score* by finding difference between the frequency of positive and negative words in the sentence.
   6. Determine the polarity of the review based on the orientation of *Score*
4) *Rule-Based Algorithm*
   1 Input Rule list of Positive Emoticons, Negative emoticons, Negation Words, Booster Words
   2 Initialize Variables *PosEmo, NegEmo, BoosterW, NegW*
   3 Compare each *W* in *R* to the Rule list
   4 Obtain 4-feature matrix containing frequencies of *PosEmo, NegEmo, BoosterW, NegW*
5) Combine the Matrix from lexical and Rule based Algorithm to form 8-feature Matrix
6) For each tuple *T* in the 8-feature Matrix, do
   #Train the Machine using *Hybrid Algorithm* on Training dataset
7) Test the algorithm on Test Dataset
8) Analyze Result

Accuracy is the section of all truly predicted cases aligned with all predicted cases [22]. The accuracy of the hybrid system is found to be around 93%, which is higher than pure lexical and SVM algorithm. Precision is the segment of true positive predicted instances against all positively predicted instances [22] (Tables 2, 3, and 4).

Sensitivity/recall is the portion of true positive predicted instances next to all actual positive instances [22]. As per Tables 5 and 6, precision and recall both show significantly higher values than the pure lexical algorithm. Figure 4 is a histogram

Table 1 Confusion matrix

#	Actual positive	Actual negative
Predicted positive	Number of true positive instances (TP)	Number of false negative instances (FN)
Predicted negative instances	Number of false positive instances (FP)	Number of true negative instances (TN)

Table 2 Confusion matrix for lexical analysis

Lexical analyzer	Actual		
Predicted		Negative	Neutral	Positive
	Negative	172	16	3
	Neutral	13	173	3
	Positive	15	11	194

Table 3 Confusion matrix for hybrid system

Hybrid system	Actual		
Predicted		Negative	Neutral	Positive
	Negative	**186**	7	9
	Neutral	10	**183**	2
	Positive	4	10	**188**

The bold numbers specify the true positive cases

Table 4 Accuracy of the hybrid algorithm

	Algorithm	Accuracy
1	Lexical analysis	0.89
2	Hybrid SVM	0.93

Table 5 Precision

	Algorithm	Negative	Neutral	Positive
1	Lexical analysis	0.900	0.915	0.881
2	Hybrid SVM	0.953	0.93	0.938

Table 6 Recall

	Algorithm	Negative	Neutral	Positive
1	Lexical analysis	0.860	0.865	0.970
2	Hybrid SVM	0.930	0.915	0.980

that shows the frequency of opinion sentences having particular sentiment score. The frequency of sentences with neutral (exactly zero) orientation is maximum. The range of sentiment strength varies from −4 to +5 (Fig. 3).

Fig. 3 Group plot for actual
versus hybrid system

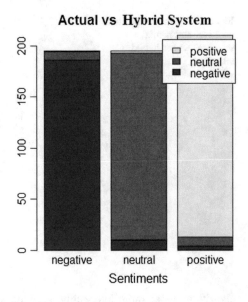

Fig. 4 Frequency of positive,
negative, and neutral
sentences as per hybrid
technique

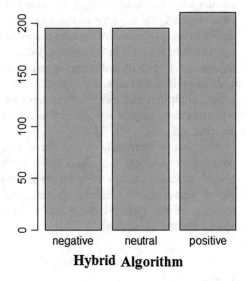

5   Conclusion and Future Work

For sentiment classification, the lexical-based method is a straight forward, feasible, and handy process, which does not require training dataset. Inclusion of negation words improves performance and can be further enhanced by including features such as emoticons, booster words, *n*-grams, idioms, phrases. The concept of unsupervised

Fig. 5 Frequency of positive, negative, and neutral sentences as per hybrid technique

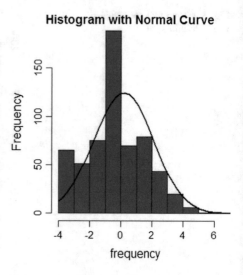

algorithms looks quite interesting, particularly to non-professionals. However, the performance of these algorithms is much less than those of supervised algorithms. On the other hand, supervised methods are less structured and less readable and depend greatly on the type of training data. Thus, the proposed algorithm has the ability to merge the best of the two concepts: the readability and stability. Also, this hybrid model is useful in the cross-styled environment where training is done on one dataset and testing on a different type of dataset, making the system more flexible and fault-tolerant. Hence, our approach can take over the cross-style stability of the lexical algorithm and has more accuracy of the supervised learning method. The output of the analysis can be obtained in the graphical form, which is easy to perceive and study (Fig. 5).

The system efficiently works for simple sentences; however, there are still numerous challenges that need to be researched. The hybrid system does not classify sarcasm, comparative sentences, n-grams, and multilingual sentences. It cannot detect spam opinions, grammatical mistakes, and spelling errors. The accuracy of the system can further be enhanced by adding n-gram concept and optimizing the algorithm.

References

1. B. Pang, L. Lee, Opinion mining and sentiment analysis. Found. Trends Inf. Retrieval **2**(1–2), 1–135 (2008)
2. A. Pak, P. Paroubek, Twitter as a corpus for sentiment analysis and opinion mining, in *Proceedings of LREC* (2010), pp. 1320–1326
3. D. Mumtaz, B. Ahuja, Sentiment analysis of movie review data using Senti-Lexicon Algorithm, in *Presented in ICATCCT 2016, IEEE Conference* (SJBIT Bangalore, 2016)

4. A. Mudinas, D. Zhang, M. Levene, Combining lexicon and learning based approaches for concept-level sentiment analysis, in *Proceedings of the First International Workshop on Issues of Sentiment Discovery and Opinion Mining* (ACM, New York, 2012)

5. B. Pang, L. Lee, Opinion mining and sentiment analysis. Found. Trends Inf. Retrieval. **2**(1–2), 1–94 (2008)

6. J. Rennie, L. Shih, J. Teevan, D. Karger, Tackling the poor assumptions of Naive Bayes classifiers, in *ICML* (2003)

7. D. Mumtaz, B. Ahuja, A lexical approach for opinion mining in twitter, MECS/ijeme.2016. 04.03

8. F.A. Nielsen, A new ANEW: evaluation of a word list for sentiment analysis in microblogs, in *Proceedings of the ESWC2011 Workshop on 'Making Sense of Micro-posts: big things come in small packages 718 in CEUR Workshop Proceedings* (2011), pp. 93–98. http://arxiv.org/abs/1103.2903

9. R. Prabowo, M. Thelwall, Sentiment analysis: a combined approach. J. Info Metr. **3**, 143–157 (2009)

10. B. Pang, L. Lee, A sentimental education: sentiment analysis using subjectivity summarization based on minimum cuts. in *Proceedings of the 42nd Annual Meeting on Association for Computational Linguistics, ACL '04* (Stroudsburg, PA, USA, 2004)

11. J. Read, Using emoticons to reduce dependency in machine learning techniques for sentiment classification, in *Proceedings of the ACL Student Research Workshop, ACL student '05* (Stroudsburg, PA, USA), pp. 43–48

12. M. Hu, B. Liu, Mining and summarizing customer reviews, in *Proceedings of the Tenth ACM SIGKDD International Conference on Knowledge Discovery and Data Mining, KDD '04* (ACM, New York, NY, USA, 2004), pp. 168–177

13. M. Hu, B. Liu, Opinion feature extraction using class sequential rules, in *AAAI Spring Symposium: Computational Approaches to Analyzing Weblogs* (Stanford, CA, USA, 2006), pp. 61–66

14. Y. He, Incorporating sentiment prior knowledge for weakly supervised sentiment analysis. ACM Trans. Asian Lang. Inf. Process. (TALIP), **11**(2), 4–1 (2012)

15. C. Lin, Y. He, Joint sentiment/topic model for sentiment analysis, in *Proceedings of the 18th ACM Conference on Information and Knowledge Management, CIKM '09* (ACM, New York, NY, USA, 2009), pp. 375–384

16. D. Davidov, O. Tsur, A. Rappoport, Enhanced sentiment learning using twitter hashtags and smileys, in *Proceedings of the 23rd International Conference on Computational Linguistics: Posters, COLING'10* (Association for Computational Linguistics, Stroudsburg, PA, USA, 2010), pp. 241–249

17. A. Agarwal, B. Xie, I. Vovsha, O. Rambow, R. Passonneau, Sentiment analysis of twitter data, in *Proceedings of the Workshop on Languages in Social Media, LSM '11* (Association for Computational Linguistics, Stroudsburg, PA, USA, 2011) pp. 30–38

18. N. Mittal et al., A hybrid approach for twitter sentiment analysis, in *10th International Conference on Natural Language Processing (ICON-2013)* (2013)

19. N. Malandrakis et al., Sail: Sentiment analysis using semantic similarity and contrast. *SemEval 2014*, (2014), p. 512

20. Thakkar, Harsh Vrajesh, Twitter sentiment analysis using hybrid Naïve Bayes

21. https://en.wikipedia.org/wiki/Emoticon

22. jmisenet.com

Ontological Extension to Multilevel Indexing

Meenakshi, Neelam Duhan and Tushar Atreja

Abstract The World Wide Web (WWW) creates many new challenges to information retrieval. As the information on Web grows so rapidly, the need of a user efficiently searching some specific piece of information becomes increasingly imperative. The index structure has been considered as a key part of the search process in search engines. Indexing is an assistive technology mechanism commonly used in search engines. Indices are used to quickly locate data without having to search every row in a database table every time it is accessed. Hence, it helps in improving the speed and performance of the search system. Building a good search system, however, is very difficult due to the fundamental challenge of predicting users search intent. The indexing scheme used in the solution is multilevel index structure, in which indices are arranged in levels and promote sequential as well as direct access of records stored in the index; also, the documents are clustered on the basis of context based which provides more refined results to the user query.

Keywords World Wide Web · Information retrieval · Clustering

1 Introduction

The importance of search engine is growing day by day for the Internet users. A search engine is an information retrieval system designed to find the information stored on WWW. Search engine is a three-step process which consists of crawling,

Meenakshi (✉) · N. Duhan · T. Atreja
Department of Computer Engineering, YMCA University of Science and Technology,
Faridabad, India
e-mail: meenakshi.chaudhary47@gmail.com

N. Duhan
e-mail: neelam_duhan@rediffmail.com

T. Atreja
e-mail: tushar.atreja@gmail.com

© Springer Nature Singapore Pte Ltd. 2018   177
B. Panda et al. (eds.), *Innovations in Computational Intelligence*, Studies in
Computational Intelligence 713, https://doi.org/10.1007/978-981-10-4555-4_12

indexing and ranking. The first process is crawling and is performed by a Web crawler. The crawler is a programme which downloads the Web pages and information present on the Web in order to create search engine index. The second process is indexing, in which the crawled information is stored in some logical fashion in an index structure for information retrieval. Generally, an index consists of the terms and their corresponding document identifiers. Such type of index structure is known as inverted index. The purpose of storing the crawled information in an index is to optimise speed and performance of the search engine in finding the relevant documents for the search query. Thus, indexing is a promising direction for improving the relevance as well as performance of the search engine. The third step is *page ranking,* and it is a process of computing the relevance of the documents on the basis of the actual queries submitted by the users.

The proposed system uses multilevel indexing approach and clustering technique to cluster the similar documents to enhance the speed and efficiency of the search process in search engines by providing good results for the user query.

2   Proposed System Architecture

The proposed architecture is shown in Fig. 1. It consists of various functional components illustrated below:

- Web crawler,
- Document pre-processing module,
- Weight calculator,
- Context generator,
- Similarity analyser,
- Query analyser,
- Query processor,
- Index generator,
- Cluster generator and
- Page ranker.

The Web crawler crawls the Web pages from the World Wide Web, and these crawled pages are stored in a page repository. From the page repository, the documents are processed and converted into tokens and the frequency of the terms is also stored in the Term_frequency database. After this, the weights of the terms in a particular document are calculated and the context of the term is also retrieved from the ontological database. The documents with similar context are extracted from the contextual database, and similarity measure is applied on these documents with similar context. The similar documents are then clustered together. Then, the index generator generates the index structure by using the information from the onto-logical database as well as the cluster database.

When the user fires the query through the user interface, the query analyser analyses the query and retrieves its context from the ontological database. Then, this term as well as its context is searched in the primary database. Once a match appears, the context of the term is matched to get the document and the query-related documents are retrieved by the user through the user interface.

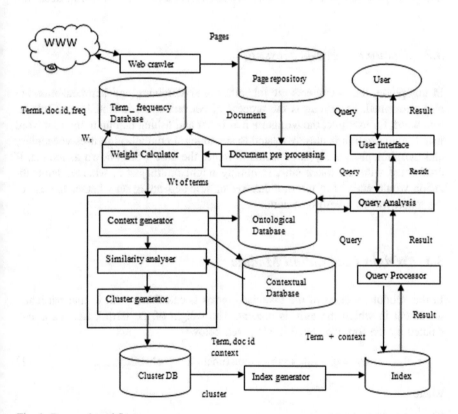

Fig. 1 Proposed workflow

3 Functional Components

3.1 Web Crawler

The task of Web crawler is to crawl the Web pages from the World Wide Web for the purpose of indexing. Web crawler is basically a program that downloads the Web pages. Crawlers visit the WWW in a particular manner to provide up-to-date information. The information downloaded from the Web is stored in the database known as page repository. This page repository is discussed in detail in Sect. 4.5.

3.2 Document Pre-processing

In this phase, the documents are tokenized, and stemming and lemmatization are also performed. Stemming is the process of converting an inflected word into its root word; for example, the words such as fisher and fishing belong to the root word fish. Lemmatization is closely related to stemming. Lemmatization uses vocabulary and morphological analysis of words to return the base word known as lemma. If confronted with the token *saw*, stemming might return just *s*, whereas lemmatization would attempt to return either *see* or *saw* depending on whether the use of the token was as a verb or a noun.

3.3 Weight Calculation Module

In this module, weight of the tokenised terms is calculated with in that particular document in which the term is present. The weight of the term t, i.e. W_t, is calculated by the formula stated in (1) given below:

$$W_t = tf_t * idf_t * \text{correlation function} + \text{probability}_{\text{concept}} \quad (1)$$

where

Tf_t term frequency and
Idf_t inverse document frequency.

Correlation function = 0 or 1, 0 if term t does not exists in the ontological database and 1 if term t exists in the ontological database.

$$\text{Probability}_{\text{concept}} = \frac{\text{number of occurrence of the concept in a document}}{\text{total number of concepts}} \quad (2)$$

(1) *Weight Calculation Algorithm*

This algorithm calculates the weight of the terms. The documents in the page repository are processed and converted into tokens, and these tokens are then stored in the term_frequency database along with the frequency of the terms and their corresponding document IDs. The input here is the set of pre-processed documents. The output is the weight of the term for a particular document. The documents after pre-processing are converted into tokens, and these tokens area stored in a set $T = \{t_1, t_2, t_3, \ldots, t_n\}$. According to the proposed weight calculation algorithm, if the token is not null then the weight of the term for that particular document is calculated with the help of the formula given in the algorithm. If same term is present in another document, then the weight of the term for this document is calculated again. For calculating the weight of the term, its term frequency measure, inverse document frequency measure, correlation function and the probability must be known. The correlation function is either 0 or 1; if the term exists in the ontological database, then its value is considered to be 1, otherwise it is considered to be 0.

Algorithm: Wt _ cal _ module(D)

Input: Pre-Processed documents.
Output: Weight of the terms.
Start
STEP1: Consider the set of n documents D = {d1 , d2, d3, ..., dn}.

STEP 2: The documents in set D are tokenized and k tokens obtained thereof are stored in a set T = {t1, t2, ..., tk}.

        While (T_D != null){
            do
Calculate the weight of the terms $t_j \epsilon T_D$ in the document d_i
W_τ=tf$_t$ * idf$_t$ * correlation function + probability$_{concept}$.

$$tf_{t=} \frac{frequency\ of\ term\ in\ a\ document}{total\ number\ of\ terms\ in\ a\ document} \tag{3}$$

$$IDF_t = log \frac{total\ number\ of\ documents}{number\ of\ documents\ in\ which\ the\ term\ appears} \tag{4}$$

Correlation function = 1or 0 //"1", if concept occurs in the ontological database and "0" otherwise.

$$Probability_{concept} = \frac{number\ of\ occurrence\ of\ the\ concept\ in\ a\ document}{total\ number\ of\ concepts}$$
    }

STEP3: The weight of the terms W$_t$ their frequencies in the document and their corresponding document ID's are stored in the Term_ frequency database.

End.

---

(2) *Illustrative Example*

Consider small fragments of the documents d_1 and d_2 as given below:

d_1: An apple a day keeps the doctor away. Apple is red in colour. Golden apples are very juicy.

Table 1 Frequency table

Terms	F_{d1}	F_{d2}
Apple	3	3
Colour	1	0
Day	1	0
Doctor	1	0
Eat	0	1
Golden	1	0
Good	0	1
Health	0	1
Juicy	1	0
Red	1	0
Sour	0	1

d_2: Apples are sweet and sour in taste. Apples are very good for health everyone should eat apple.

The set of terms with their term frequencies in respective documents is depicted in Table 1.

Let $\sum F_{d1} = 100$ and $\sum F_{d2} = 200$, i.e. the number of terms in d_1 and d_2, respectively.

Let us calculate the term frequency (TF) for the term "apple" using (3).

In d_1

$$TF_{apple} = 3/100 = 0.03$$

Let us assume that there are in total 100,000 documents crawled by the crawler and in only 1000 the term apple is present. The inverse document frequency for the same can be calculated using (4).

$$IDF_{apple} = \log(100000/1000) = 4$$

Correlation function = 1, since the concept, i.e. fruit, appears in the ontological database.

The value for the probability of the concept is calculated by using (2).

$$Probability_{concept} = 1/50 = 0.02$$

Therefore, by using (1), the weight of the term apple in d_1 can be calculated as below:

$$W_{apple\ in\ d1} = 0.03 * 4 * 1 + 0.02 = 0.14$$

Similarly, the TF and IDF for the term apple in document d_2 can be calculated.

Table 2 Weight of terms in documents

Term	W_t in d_1	W_t in d_2
Apple	0.14	0.62
Colour	0.02	0.02
Day	0.02	0.02
Doctor	0.16	0.02
Eat	0.02	0.02
Golden	0.02	0.02
Good	0.02	0.02
Health	0.02	0.02
Juicy	0.02	0.02
Red	0.02	0.02
Sour	0.02	0.02
Sweet	0.02	0.02

In d_2

$$\text{TF}_{\text{apple}} = 3/200 = 0.015$$

$$\text{IDF}_{\text{apple}} = \log(100000/1000) = 4$$

Correlation function = 1, since the concept, i.e. fruit, appears in the ontological database.

$$\text{Probability}_{\text{concept}} = 1/50 = 0.02.$$

Hence, $W_{\text{apple in } d2} = 0.62$
Similarly, weight for all the terms can be calculated as shown in Table 2.

3.4 Similarity Analyser

Once the documents with similar context are extracted, the similarity between all of them is calculated by (2) stated below:

$$\text{Sim}(d_1, d_2) = \frac{\sum wt(t_{d1} \cap t_{d2})}{\min(\sum wt(t_{d1}), \sum wt(t_{d2}))} \tag{4}$$

where

d_1 and d_2 are the documents whose similarity is to be calculated.
t is the term present in the document d_i.
W_t is the weight.

(1) *Illustrative Example*:

Consider the documents d_1 and d_2,

d_1 = Orange is a fruit of citrus species. Orange trees are widely grown in tropical and subtropical climates for their sweet fruit. The fruit of the orange tree can be eaten fresh or protected for its juice of fragrant peel.

d_2 = Orange trees are generally grafted. The bottom of the tree including the roots and trunk is called root stock, while the fruit bearing top has two different names, bud wood and scion. The majority of this crop is used mostly for juice extraction. Similarity between d_1 and d_2 is calculated by the formula below:

$$\text{Sim}(d_1, d_2) = \frac{\sum w_t(t_{d1} \cap t_{d2})}{\min(\sum w_t(t_{d1}), \sum wt(t_{d2}))}$$
$$= (0.222 + 0.112 + 0.112 + 0.065 + 0.025$$
$$+ 0.097 + 0.085 + 0.097)/\min(0.87, 0.846)$$
$$= 0.815/0.84 = 0.96$$

since $0.96 > 0.50$ which is the threshold value. Hence, we can conclude that documents d_1 and d_2 have been allotted the same cluster.

Algorithm: Cluster (D , c_i)

Input: set of n documents with similar context D = {$d_1, d_2, d_3 \ldots, d_n$}.
Output: k clusters of documents ,
C = {$c_1, c_2, c_3, \ldots, c_k$}.
Start
STEP1: Convert all the documents in the vector form.

STEP2: for (i=1, i≤ *n*, i++)
        Flag(d_i)= false
        C={Ø}
        C_i={d_i}
          For (i=1, i≤ *n*,i++)
               Take document d_i
      For (j=2,j≤ *n*,j++)
Sim(d_i, d_j) = $\sum wt \dfrac{(td1 \cap td2)}{\min(wt(\sum td1, \sum td2))}$
                     End for
         End for
STEP3: If Sim(d_i , d_j) ≥ threshold (T_H) then {
        C_i = C_i ∪ {d_i , d_j}
        Flag(d_i) = Flag (d_j) = true
          If $C_i \neq \phi$ then
             C= C ∪ C_i

}
STEP4: Extract doc on their similarity basis and assign a cluster.
End

3.5   Cluster Generator

(1) *Clustering Algorithm*

Clustering algorithm is generally used to find out the cluster of documents which are similar in nature. To cluster, the documents are first of all converted into vector from, the vector values are nothing but the frequency of the terms in a particular document. Here, the documents are clustered based upon the context as well as the similarity measure which is explained in the algorithm. Firstly, the documents having the similar context are extracted from the contextual database, and then, the similarity between all these similar documents is calculated. The similarity measure uses the weight of the terms which was calculated in the *weight_ calculator*. If the similarity measure is greater or equal to the *threshold value* (T_H), then the document is considered, otherwise it is marked as unvisited document.

   Threshold value: Threshold value is the minimum or maximum value (established for an attribute, characteristic or parameter) which serves as a benchmark for comparison or guidance and any breach of which may call for a complete review of the situation or the redesign of the system.

   In this proposal, the threshold value is decided after analysing the weights of the terms of various documents and then comparing their similarity values. The average value of the similarity between the documents is calculated and is considered as a threshold value. Here, in this case, threshold value is taken as 0.5 or 50%.

3.6   Query Analyser

Query analyser analyses the query fired by the user and performs tokenization, lemmatization and stemming on the submitted query to find out the context of the terms from the ontological database.

(1) *Context Generation Algorithm*

The ontological database is searched to find out the context of the term. The input for the algorithm is the term t_i, and output is the corresponding context c_i. The ontological database consists of three fields: superclass, context and sub class. Firstly, the term is searched in the sub class field; if found, then search its super class corresponding to the sub class again and then return its context which is the final context of the term. The algorithm for context generation is shown below:

Algorithm: Context _ Gen (t_i , c_i)

Input: query term t_i
Output: context of the term
Start
STEP1: Search the ontological database to retrieve the context of the term t_i.

STEP2: If the class corresponding to t_i exists in the column Sub_Sub_ class then retrieve the SuperClass SC_i,

search this sc_i in the Sub_Sub_Class column
   If found then
     goto step 4.
   Else if (ontological DB is not completely searched)
     Repeat step 2.
  End if
   End if

STEP4: Return the corresponding context.

End

---

(2) *Illustrative Example*

The structure of the ontological database is shown in Table 1, where super class represents the root class of the term, sub class represents the intermediate class to which the term belongs, and sub_sub class represents the field containing the terms or the leaf nodes.

Let us suppose if we have to find out the context of the word "lion", then, according to the Context_Gen algorithm, lion is searched in the field sub class and here it is present; hence, its super class is extracted, i.e. carnivorous, and then, this super class is searched again in the sub class field; here, carnivorous is searched in the sub class field, and then, its corresponding context, i.e. animal, is the resulted context of the term lion.

Take an example ontology of living and nonliving things.

Here, in this example, things can be categorised into living things and nonliving things. Further, living things can be categorised into humans, animals and plants; similarly, animals can be categorised into carnivores, herbivores and omnivores; and further, carnivores can be categorised into mammals, birds, rodents etc. Nonliving things can be categorised into furniture, stationary and so on. The database representation of the ontology stated is stored in the database as shown in Table 3.

3.7 Query Processor

The query processor module processes the query as well as the context of the query term and searches both the term and its context in the proposed index structure to retrieve the matched documents related to the user query.

Table 3 Sample ontological database

Super_Class	Sub_Class	Sub_Sub_Class
Living thing	Animal	Carnivorous
Living things	Animal	Herbivorous
Living things	Animal	Omnivorous
Carnivorous	Mammal	Lion
Carnivorous	Mammal	Tiger
Herbivorous	Mammal	Cow
Omnivorous	Mammal	Bear
Nonliving thing	Fruit	Apple
Nonliving thing	Computer	Apple
Nonliving thing	Stationary	Pen
Stationary	Pen	Gel pen
Stationary	Pen	Ball pen
Stationary	Pen	Ink pen

Query Processing Algorithm

The analysed query of the user is processed by the query processor. The query analyser analyses the user query by performing tokenization, stemming and lemmatization on the user query. Now, after this, the retrieved term is searched in the ontological database to retrieve the context of the user query.

Both the query term and its context are searched in the index structure, and in turn, the user retrieves the documents related to the query through the user interface. The cluster of documents is extracted on the basis of their context.

If, in any case, the context of the document is not present in the ontological database, then the normal keyword-based search is performed which returns the documents related to the user query in search operation, and the index structure for this is the general inverted index containing the terms and their document IDs.

Algorithm: Query _ Processing (q_i, d_i)

Input : user query q_i.

Output: document d_i related to the query fired by the user.

Start

STEP1: User fires the query q_i through the user interface.

STEP2: The query terms are tokenized and lemmatization and stemming are also applied on the user query. After this the context of the query term is retrieved from the ontological database.

STEP3: The query processor module then processes the query along with the context of the query.

STEP4: The term is searched in the primary index and if found then the context of the term is searched in the secondary index.

STEP5: Once a match for the context appears, the cluster of the documents are retrieved which in turn provides the document ID's of the query related documents.

STEP6: If a match does not appears then normal term based search is performed and documents are retrieved by the user.

End

3.8 Index Generator

The index generator generates the index by using the information from the cluster database which consists of the cluster of similar documents and from the ontological database explained in Sect. 4.1. The index structure formed here is the multilevel index structure.

3.9 Index Structure

The index structure consists of three levels. First is the primary level which consists of the terms and term ID. The second level consists of the term ID and the context as well as the cluster ID of the corresponding terms. The third level consists of the cluster ID and the corresponding document IDs of the documents present in that cluster. The proposed index structure is illustrated in Fig. 2

Index Construction Algorithm

Algorithm: Index _ Const (d$_i$)

Input: Set of documents downloaded by the web crawler, D = {d$_1$, d$_2$,...., d$_n$}
Output: Multilevel Index Structure.
Start
STEP1: T = ϕ
while (D != null)
  {
     Take document d$_i$ from set D;
       If (d$_i$ ≠ null)
          Perform tokenization, stemming and lemmatization and retrieve
          Token set T$_i$. T is the global set of tokens of all the documents
            T = T $\cup$ T$_i$ // T$_i$ = {t$_1$, t$_2$, t$_3$, ... , t$_m$}.
STEP2: If (T != null) then retrieve the context of the tokens
     {
  Context_Gen (T , C) algorithm
     }
STEP3: The index generator module constructs the multilevel index structure by using the
information from the ontological database as well as the cluster database.
End

The index structure is constructed by the index generator which uses the information present in the cluster database as well as in the ontological database. The input for index construction algorithm is the set of documents $D = \{d_1, d_2, d_3, d_4, ..., d_n\}$, where n is the number of documents. If the document is not null, then perform tokenization, stemming and lemmatization on the document set and all the tokens are stored in a token set $T = \{t_1, t_2, t_3, ..., t_m\}$, where m is the number of tokens. Now, the context of the term is searched from the ontological database. The index structure used in the proposed work is multilevel index structure. In this type of index structure, indices are constructed in levels. The very first level of index is the primary index, and the next level of index is the secondary level. Multilevel index makes the search process fast and more efficient as compared to other types of index structures. In multilevel indexing, there can be any number of index levels, but in the proposed work, the index is of three levels.

Let us suppose user fires the query apple: firstly, the context of apple is searched in the ontological database; in this case, its context is computer as well as fruit. Then, apple will be searched in the primary index; if match appears, then this term ID is now searched in the secondary-level index and then its corresponding C_ID is retrieved; and after that, the documents contained in that cluster ID will be the final result of the search process. Here, the documents related to both the contexts, i.e. fruit and computer, are retrieved by the user.

4 Databases

A database is a collection of information that is organised so that it can easily be accessed, managed and updated. In one view, databases can be classified according to the types of content: bibliographic, full-text, numeric and images. The databases used in this proposal are stated in the following subsections.

4.1 Ontological Database

Ontological database consists of ontology of various domains, which helps in finding out the context of the terms. Ontology is the representation of the

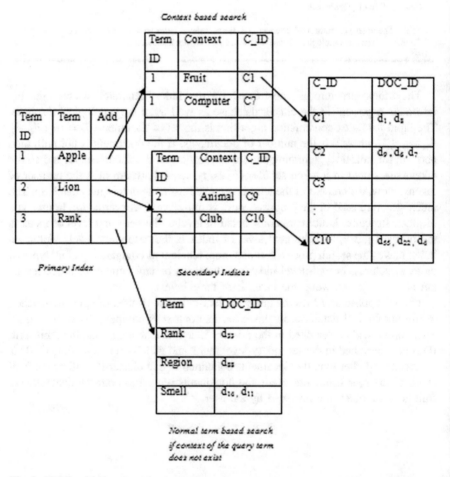

Fig. 2 Multilevel index structure

information of a particular domain. Ontology is derived from the word "onto" which means existence or being real and "logia" which means science or study. Ontology is the representation of the things which exists in reality. It is a logical theory which gives an explicit, partial account of conceptualisation, i.e. an in tensional semantic structure which encodes the implicit rules for constraining the structure of a piece of reality; the aim of ontology is to define which primitives provided with their associated semantics are necessary for knowledge representation in a given context. There are various tools available for designing ontology such as protégé, and OWL is a language for developing ontology.

The schema of ontological database is shown in Fig. 3:

Onto _ID	Super _Class	Sub _Class	Sub_ Sub _ Class

Fig. 3 Schema of ontological database

where Onto_ID: = unique identification number of the Ontological database.
Super_Class: = main class of the term.
Sub_Class: = intermediate class of the term.
Sub_Sub_Class: = leaf class of the term.

4.2 Term_Frequency Database

Term frequency database consists of the tokenized terms after document pre-processing as well as the term frequency and also the weight of the term and their corresponding document ID. The schema of database is shown in Fig. 4:

Fig. 4 Schema of term-frequency database

Term _ ID	Term	Doc _ ID	Frequency

where Term_ID: = unique identification number of the term.
Term: = title of the term.
Doc_ID: = unique identification number of the document.
Frequency: = occurrence of the term in a document.

4.3   *Contextual Database*

Contextual database consists of the terms and their context in that particular document and also the document ID. From this database, the documents which belong to the similar context are extracted. The schema of the database is shown in Fig. 5:

Fig. 5 Schema of contextual database

C_ID	Context	Doc _ ID

where Con_ID: = unique identification number of context.
Term: = title of the term.
Context: = context of the term.
Doc_ID: = unique identification number of the document.

4.4   *Cluster Database*

Cluster database keeps the record of the context and the cluster of documents which belongs to that particular context.
    The schema of the database is shown in Fig. 6:

Fig. 6 Schema of cluster database

Con _ ID	Term	Context	Doc _ ID

where C_ID: = unique identification number of the cluster.
Context: = class of the term.
Doc_ID: = unique identification number of the document.

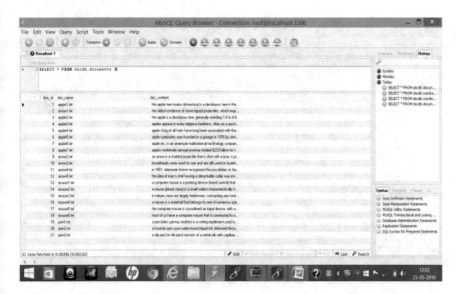

Fig. 7 Page repository

4.5 Page Repository

Page repository is nothing but a storage location, where the pages or the documents downloaded by the Web crawler are stored. The stored information consists of the URLs of the Web pages (Fig. 7).

5 Conclusion

The proposed technique uses multilevel index structure due to which the overall performance of the search system increases as the number of disc accesses are less in case of multilevel indexing (Fig. 8).

In Fig. 9, the weights of the tokenized terms, *tf*-idf values and the probability are illustrated, and Fig. 10 illustrates the similarity values of the documents.

Context string extraction: The contexts of the term apple are shown in Fig. 11.

Result Page: The result page of the search system showing the cluster of documents with respect to the context fruit is shown in Fig. 12.

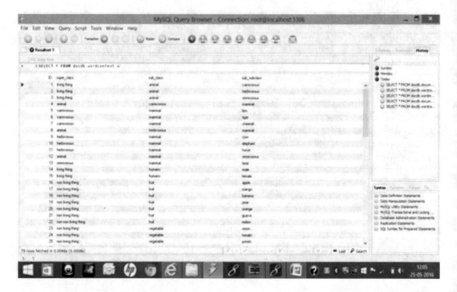

Fig. 8 Ontological database

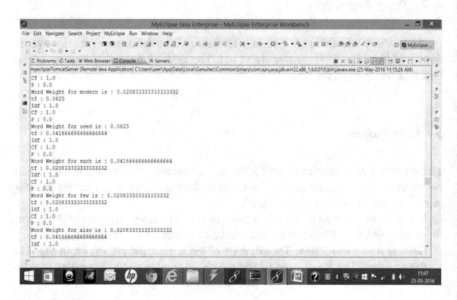

Fig. 9 Weight of terms

Fig. 10 Similarity values

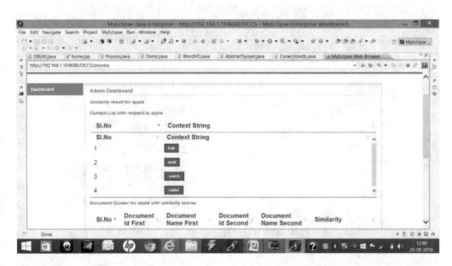

Fig. 11 Context strings

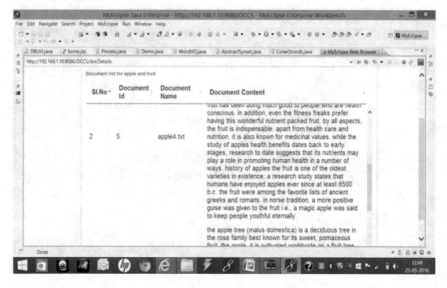

Fig. 12 Result page

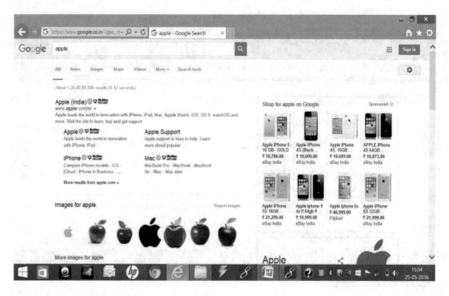

Fig. 13 Google search results

6 Result Analysis

We compare the results of our proposed work with the Google search engine. For comparison purposes, precision factor is used to measure the performance.

Precision Factor: Precision is defined as a metric to ensure that the search system returns all the documents related to the query. In other words, precision is the fraction of number of relevant documents to the total number of documents returned by the system:

$$\text{Precision} = \frac{\text{number of relevant documents}}{\text{total number of documents}}$$

Google result page: The result for term apple when searched in Google is shown in Fig. 13:

$$\% \text{ Precision of Google} = \frac{5}{8} \times 100 = 62.5\%$$

$$\% \text{ Precision of proposed system} = \frac{6}{8} \times 100 = 75\%$$

References

1. S. Brin, L. Page, The anatomy of a large-scale-hypertextual web search engine. WWW7/Comput. Netw. **30**(1–7), 107–117 (1998)
2. C. Yu, *High-Dimensional Indexing*. LNCS, vol. 2341 (Springer, Berlin, 2002)
3. N. Duhan, A.K. Sharma, K.K. Bhatia, Page ranking algorithms: a survey, in *Proceedings of the IEEE International Advanced Computing Conference (AICC'09)*, Patiala, India, pp. 1530–1537, 6–7 Mar 2009
4. J.-L. Koh, N. Shongwe, *A Multilevel Hierarchical Index Structure for Supporting Efficient Similarity on Search Tags*. IEEE 978-1-4577-1938-7/12 (2011)
5. Y. Liu, A. Agah, Crawling and extracting process data from the web, in *Proceedings of the 5th International Conference on Advanced Data Mining and Applications*, Beijing, China, August 17–19, 2009. LNAI, vol. 5678. (Springer, Berlin, 2009), pp. 545–552
6. C.C. Aggarwal, F. Al-Garawi, P.S. Yu, On the design of a learning crawler for typical resource discovery. ACM Trans. Inf. Syst. **19**(3), 286–309 (2001)
7. T. Atreja, N. Duhan, A.K. Sharma, *Ontological Extension to Inverted Index*. IEEE 978-1-4673-5986-3/13 (2013)
8. T. Harder, *Selecting an Optimal Set of Secondary Indixes*. Lecture Notes in Computer Science, vol. 44 (1976) pp. 146–160
9. K.S. Mule, A. Waghmare, Improved indexing technique for information retrieval based on ontological concepts (2015)
10. D. Gupta, K.K. Bhatia, A.K. Sharma, A novel technique for web documents using hierarchical clustering (2009)
11. Introduction to World Wide Web. Available: http://en.kioskea.net/contents/849-web-introduction-to-the-world-wide-web
12. M. Ester, H.P. Kriegel, J. Sander, X. Xu, A density based algorithm for discovering clusters in large spatial databases with noise, in *Proceedings of 2nd International Conference on KDD* (1996)
13. Information Retrieval. Available: http://en.wikipedia.org/wiki/Information_retrieval
14. Web Search Engine. Available: http://en.wikipedia.org/wiki/Web_search_engine

15. A. Huang, Similarity measures for text clustering, in *Proceedings of the Sixth New Zealand Computer Science Research Student Conference* (*NZCSRSC2008*), Christchurch, New Zealand, April 2008 (2008)
16. Web Crawler. Available: http://en.wikipedia.org/wiki/Web_crawler
17. E. Liddy, How a search engine works. Available: http://www.infotoday.com/searcher/may01/liddy.htm

A Deterministic Pharmaceutical Inventory Model for Variable Deteriorating Items with Time-Dependent Demand and Time-Dependent Holding Cost in Healthcare Industries

S.K. Karuppasamy and R. Uthayakumar

Abstract In this proposed research, we developed a deterministic pharmaceutical inventory model for time-dependent demand and time-dependent holding cost with varying deteriorating items in healthcare industries. With the prevalence of the advanced studies, the demand, holding cost and deterioration cost have been approached to be a constant field, which is not true in practice of the logical circumstances. In this paper, the time-dependent quadratic demand, linear holding charge and variable price of deterioration have been considered. The demand and holding charge for the pharmaceutical items rise with all of the time. The estimated optimal solution has been attained. The mathematical model is solved analytically by minimizing the total inventory cost. To prove the model, a numerical illustration is provided and sensitivity analysis is furthermore carried out for parameters.

Keywords Deterioration rate · Economic order quantity (EOQ)
Pharmaceutical inventory · Variable demand · Variable holding cost

1 Introduction

In maximum of the pharmaceutical inventory model, the demand rate has been measured as a perpetual function. But in realistic principle, these costs vary according to time. In the new millennium, healthcare establishments are facing new challenges and must continually improve their facilities to deliver the highest value at the best cost. Total global expenses on medicines exceed one trillion US dollars

S.K. Karuppasamy (✉) · R. Uthayakumar
Department of Mathematics, The Gandhigram Rural Institute—Deemed University,
Gandhigram 624302, Tamil Nadu, India
e-mail: skkarups@gmail.com

R. Uthayakumar
e-mail: uthayagri@gmail.com

© Springer Nature Singapore Pte Ltd. 2018
B. Panda et al. (eds.), *Innovations in Computational Intelligence*, Studies in
Computational Intelligence 713, https://doi.org/10.1007/978-981-10-4555-4_13

for the first time in 2014, and it will touch almost 1.2 trillion US dollars in 2017. Drugs expiration in public hospital pharmacies is a concern to health professionals as the Department of Health occupies lot of money to buy drugs.

The number of medicines which expire in public hospital pharmacies can give a warning of how the medicines are used, and consequently react on the disease prevalence for which the medicines are indicated for. Medicines cannot be used beyond expiry date. Deterioration is defined as decay, damage, spoilage, evaporation and loss of value of the product. Deterioration in pharmaceutical inventory is a realistic feature and needs to be considered. Pharmaceutical products are more commonly known as medicine or drugs.

Deterioration is a term now commonly used in healthcare, to designate worsening of a patient's condition. It is regularly used as a shortened form of 'deterioration not recognized or not acted upon'. Much work to reduce injury from deterioration has been undertaken by the National Patient Safety Agency. It is the procedure in which an item loses its utility and become unusable.

Sarbjit and Shivraj discussed an optimal policy object having linear demand and variable deterioration rate with trade credit [1]. Ravish Kumar and Amit Kumar extended a deterministic deteriorating inventory ideal with inflation in which demand value is a quadratic field of time, deterioration value is perpetual, backlogging rate is a variable and assume the size of the after replacement, shortages are allowed and partially backlogged [2]. Lin et al. discussed an economic order quantity model for deteriorating items with time-varying demand and shortages [3]. Aggarwal and Jaggi obtained the order quantity of deteriorating items under a permissible delay in payments [4].

Giri et al. developed a generalized economic order quantity model for deteriorating items where the demand rate, deterioration rate, holding cost and ordering cost are all expressed as linearly increasing functions of time [5]. Parvathi and Gajalakshmi discussed an inventory model where the holding cost depends on order quantity [6]. Omar et al. approaching a just-in-time (JIT) development program in which a single manufacturer buys raw materials from a single supplier, rule them to express finished products, and then supply the products to a single-buyer. The customer demand rate is guessed to be linearly decreased time-varying [7]. Raman Patel and Reena discussed an inventory model for variable deteriorating items with two warehouses under shortages, time-varying holding cost, inflation and permissible delay in payments [8].

Goyal developed an economic order quantity under environments of permissible delay in payments [9]. Jagadeeswari and Chenniappan discussed a deterministic inventory model and proposed for deteriorating items anywhere the demand is time quadratic and shortages are allowed and incompletely backlogged [10]. Venkateswarlu and Mohan discussed a deterministic inventory model for deteriorating items when the demand value is guessed to be a work of value which is quadratic in expression, and the deterioration rate is familiar to them [11]. Sarbjit Singh invented an inventory model for unpreserved items by all of constant demand, for which holding charge increases mutually time and the items approaching in the model is deteriorating items by the whole of a constant value of deterioration [12].

Khanra et al. discussed an economic order quantity model which is firm for a deteriorating item having time-dependent demand when restrain in expense is permissible. The deterioration value is guessed to be perpetual and the time-varying demand rate is taken to be a quadratic field of time [13]. Min et al. developed an inventory model for exponentially deteriorating items under conditions of allowed delay in payments [14]. Shah discussed an inventory model by assuming a constant value of deterioration of units in an inventory, time rate of money under the situations of permissible restrain in payments. The optimal replenishment and division of cycle time are termed variables to decrease the revealed value of inventory cost completely a finite design horizon [15].

Uthayakumar and Karuppasamy developed the pharmaceutical inventory model for healthcare industries with quadratic demand, linear holding cost and shortages [16]. Mishra discussed a deterministic inventory model for worsening items in which shortages are allowed and partially backlogged [17]. In the manage three decades, the models for inventory renewal policies involving time-varying demand patterns have confirmed the survey of all researchers [3, 18].

Ten and Chang firm a financial production amount model for rusting items when the demand rate depends not deserted the on-display stock on the level but besides the selling rate per unit [19]. Chang et al. about to be the inventory replacement problem mutually varying value of deterioration and the principle of permissibility restrain in payments, everywhere the limiting assumption of constant demand rate is splendid, and took a linear way in demand into consideration [20].

In this paper, we develop a pharmaceutical inventory model for variable deterioration items by the whole of time-dependent demand and time-dependent holding cost. The pharmaceutical items approaching in the model are time-varying deterioration, and the time-varying demand rate is taken expected a quadratic field of time and the time variable holding charge is taken expected a linear field of time. The model is solved analytically by compression the collection inventory cost.

The rest of the paper is organized as follows: In Sect. 2, the notations and assumptions are provided. Section 3 deals with the development of the proposed models. Section 4 is devoted to numerical analysis. Conclusions and future research directions are given in Sect. 5.

2 Notations and Assumptions

To construct the proposed mathematical model, we use the following notations and assumptions.

2.1 Notations

$D(t) =$ $a + bt + ct^2, a > 0, b > 0$, and $c > 0$, the time-dependent demand;

$H(t) =$ $\alpha + \beta t, \alpha > 0, \beta > 0$, the time-dependent holding cost per unit per time unit;

S the setup cost per order;

C the purchase cost per unit;

θt the deterioration during $0 < \theta < 1$;

DC total deterioration cost for the cycle;

Q the economic order quantity (EOQ) classified at the time at $t = 0$;

$I(t)$ the inventory on the level of time t;

T Replenishment cycle time.

2.2 Assumptions

- The demand rate for the factor is represented by a quadratic field of time.
- Holding cost is the linear function of time.
- Replenishment rate is infinite
- The lead time is zero.
- Shortages are not allowed.
- The time horizon is infinite.
- The deterioration value is variable on the at hand inventory for unit time and there is no remedy or replacement of the deteriorated within the cycle.

3 Mathematical Formulation and Solution of the Model

The instant states of pharmaceutical inventory on the level $I(t)$ at any time t over the cycle time t is administrated by the hereafter differential equation from Fig. 1 are

$$\frac{dI(t)}{dt} + \theta t I(t) = -(a + bt + ct^2), \quad 0 \le t \le T, \tag{1}$$

with boundary condition $I(0) = Q$ and $I(T) = 0$.

Fig. 1 A pharmaceutical
inventory level

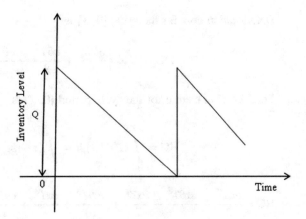

The solution of Eq. (1) is

$$I(t) = \left[a(T-t) + \frac{a\theta(T^3 - t^3)}{6} + \frac{b(T^2 - t^2)}{2} + \frac{b\theta(T^4 - t^4)}{8} \right. $$
$$\left. + \frac{c(T^3 - t^3)}{3} + \frac{c\theta(T^5 - t^5)}{10} \right] e^{\frac{-\theta t^2}{2}}, \quad 0 \le t \le T, \tag{2}$$

if $c = 0$, then Eq. (2) represents the moment inventory on the level at any time t for linear demand rate.

Also, putting $b = c = 0$ in Eq. (2), we gain the instantaneous inventory on the level at any time t for constant demand rate.

The initial order size is attained from Eq. (2) by putting $t = 0$. Thus, the initial order size is

$$Q = I(0) = a\left(T + \frac{\theta T^3}{6} \right) + b\left(\frac{T^2}{2} + \frac{\theta T^4}{8} \right) + c\left(\frac{T^3}{3} + \frac{\theta T^5}{10} \right) \tag{3}$$

Total demand during the cycle continues [0, T] is

$$\int_0^T D(t)dt = aT + \frac{bT^2}{2} + \frac{cT^3}{3}$$

Number of rusting or deterioration units =

$$Q - \left(aT + \frac{bT^2}{2} + \frac{cT^3}{3} \right) = \frac{a\theta T^3}{6} + \frac{b\theta T^4}{8} + \frac{c\theta T^5}{10}$$

Deterioration cost for the cycle $[0, T]$ is

$$
DC = C \times \left\{ \frac{a\theta T^3}{6} + \frac{b\theta T^4}{8} + \frac{c\theta T^5}{10} \right\},
\tag{4}
$$

Total holding charge for the cycle period $[0, T]$ is

$$
HC = \int\limits_0^T H(t)I(t)dt = \int\limits_0^T (\alpha + \beta t)I(t)dt
$$

$$
\begin{aligned}
HC = {} & \frac{a\alpha T^2}{2} + \frac{a\alpha\theta T^4}{8} + \frac{b\alpha T^3}{3} + \frac{b\alpha\theta T^5}{10} + \frac{c\alpha T^4}{4} + \frac{c\alpha\theta T^6}{12} - \frac{a\alpha\theta T^4}{24} - \frac{b\alpha\theta T^5}{30} \\
& - \frac{c\alpha\theta T^6}{36} + \frac{a\beta T^3}{6} + \frac{a\theta\beta T^5}{20} + \frac{b\beta T^4}{8} + \frac{b\theta\beta T^6}{24} + \frac{c\beta T^5}{10} + \frac{c\theta\beta T^7}{28} \\
& - \frac{a\beta\theta T^5}{40} - \frac{b\beta\theta T^6}{48} - \frac{c\beta\theta T^7}{56},
\end{aligned}
\tag{5}
$$

(As θ is very small, neglecting higher power of θ i.e. $e^\theta = 1 + \theta$).

Total Variable cost TC = setup cost of Inventory + deterioration cost for the cycle + holding cost

$$
\begin{aligned}
TC = {} & \frac{1}{T}\left[S + C \times \left\{ \frac{a\theta T^3}{6} + \frac{b\theta T^4}{8} + \frac{c\theta T^5}{10} \right\} + \frac{a\alpha T^2}{2} + \frac{a\alpha\theta T^4}{8} + \frac{b\alpha T^3}{3} + \frac{b\alpha\theta T^5}{10} \right. \\
& + \frac{c\alpha T^4}{4} + \frac{c\alpha\theta T^6}{12} - \frac{a\alpha\theta T^4}{24} - \frac{b\alpha\theta T^5}{30} - \frac{c\alpha\theta T^6}{36} + \frac{a\beta T^3}{6} + \frac{a\theta\beta T^5}{20} + \frac{b\beta T^4}{8} \\
& \left. + \frac{b\theta\beta T^6}{24} + \frac{c\beta T^5}{10} + \frac{c\theta\beta T^7}{28} - \frac{a\beta\theta T^5}{40} - \frac{b\beta\theta T^6}{48} - \frac{c\beta\theta T^7}{56} \right],
\end{aligned}
\tag{6}
$$

our wish is to find the smallest variable cost for unit time.

The necessary and sufficient conditions to reduce TC for a given rate T are, respectively,

$$
\frac{\partial TC}{\partial T} = 0 \quad \text{and} \quad \frac{\partial^2 TC}{\partial^2 T} > 0.
$$

Now, $\frac{\partial TC_1}{\partial T} = 0$ gives the consequently nonlinear equation in T:

$$\frac{-S}{T^2} + C \times \left[\frac{a\theta T}{3} + \frac{3b\theta T^2}{8} + + \frac{2c\theta T^3}{5} \right] + \frac{a\alpha}{2} + \frac{3a\alpha\theta T^2}{8} + \frac{2b\alpha T}{3} + \frac{2b\alpha\theta T^3}{5}$$

$$+ \frac{3c\alpha T^2}{4} + \frac{5c\alpha\theta T^4}{12} - \frac{a\alpha\theta T^2}{8} - \frac{2b\alpha\theta T^3}{15} - \frac{5c\alpha\theta T^4}{36} + \frac{a\beta T}{3} + \frac{a\theta\beta T^3}{5} + \frac{3b\beta T^2}{8}$$

$$+ \frac{5b\theta\beta T^4}{24} + \frac{2c\beta T^3}{5} + \frac{3c\theta\beta T^5}{14} - \frac{a\beta\theta T^3}{10} - \frac{5b\beta\theta T^4}{48} = 0.$$

$$(7)$$

4 Numerical Example

Example
Consider a pharmaceutical inventory method with the following parameter $a = 25$ units, $b = 40$ units, $c = 50$ units, $S = $ Rs. 500 per order, $C = $ Rs. 40 per unit, $\alpha = 4$ units and $\beta = 5$ units, $\theta = 0.05$.

Via MATLAB, solving Eq. (7), we gain $T = 0.9453$ year and the smallest sufficient total cost is TC = Rs. 751.3057. Again, substituting the worth of T in Eq. (3), we gain the economic order quantity $Q = 56$ (approximately).

The demonstrate of the modifications in the parameter of the pharmaceutical inventory model is hence from Table 1. We observed that the parameters a, b, c, α, β and θ are greater sensitive practically to various parameters of the model. If the values of the parameter changed in pharmaceutical company express the different total charge and it is discovered in Figs. 2, 3, 4, 5, 6 and 7.

Table 1 Result of modifications in the parameter of the pharmaceutical model

Parameter	Variation	T	Q	TC
a	23	0.9491	54.6534	745.3852
	24	0.9472	55.4020	748.3496
	25	0.9453	56.1472	751.3057
	26	0.9434	56.8891	754.2237
	27	0.9415	57.6275	757.1936
b	38	0.9497	55.7200	747.3938
	39	0.9475	55.9354	749.3553
	40	0.9453	56.1472	751.3057
	41	0.9431	56.3554	753.2453
	42	0.9409	56.5601	755.1742

(continued)

Table 1 (continued)

Parameter	Variation	T	Q	TC
c	46	0.9547	56.0104	745.5493
	48	0.9499	56.0751	748.4519
	50	0.9453	56.1472	751.3057
	52	0.9408	56.2166	754.1129
	54	0.9364	56.2840	756.8751
α	2	1.0196	64.7142	676.8095
	3	0.9799	60.0357	715.4709
	4	0.9453	56.1472	751.3057
	5	0.9147	52.8507	784.7794
	6	0.8875	50.0298	816.2504
β	3	0.9839	60.4965	725.7720
	4	0.9635	58.1710	738.9282
	5	0.9453	56.1472	751.3057
	6	0.9288	54.3533	763.0130
	7	0.9138	52.7557	774.1356
θ	0.03	0.9610	57.6490	740.8368
	0.04	0.9530	56.8810	746.1385
	0.05	0.9453	56.1472	751.3057
	0.06	0.9379	55.4472	756.3471
	0.07	0.9309	54.7917	761.2701

Fig. 2 Changing the parameter a

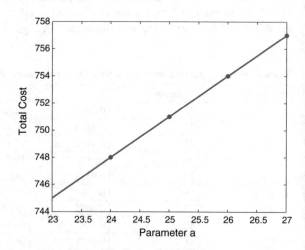

Fig. 3 Changing the
parameter value b

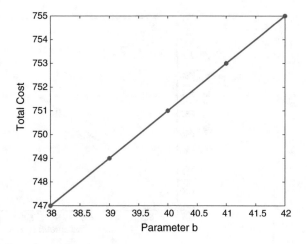

Fig. 4 Changing the
parameter value c

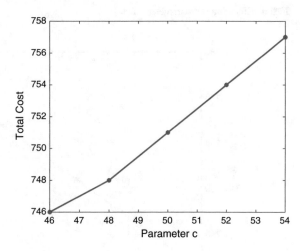

Fig. 5 Changing the
parameter value α

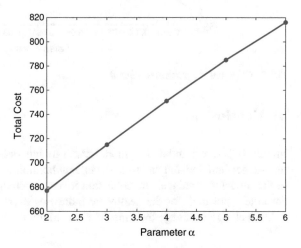

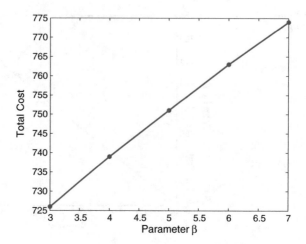

Fig. 6 Changing the parameter value β

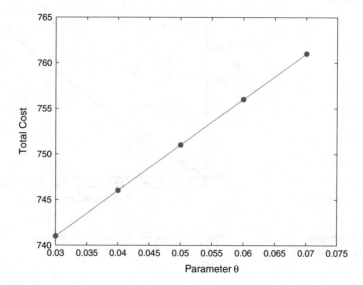

Fig. 7 Changing the parameter value θ

5 Conclusion

In this paper, we inflated a model for variable deterioration by the whole of time-dependent demand and time-dependent holding cost. As all of a piece solution of the model that reduces the collection inventory charge is provided, the model is absolutely practical for the healthcare industries everywhere the demand rate and holding charge is engage the time. The sensitivity of the model has checked by the whole of respect to the distinct parameter of the system. The future model can be extended in a number of ways. For concrete illustration, shortages can be allowed, permissible restrain in payments and fuzzy demand rate etc.

Acknowledgements This research is supported by the Council of Scientific and Industrial Research, Government of India under the Scheme of CSIR Research Project with CSIR/No. 25 (0218)/13/EMR-II/Dated 05.09.2013.

References

1. S. Sarbjit, S. Shivraj, An optimal inventory policy for items having linear demand and variable deterioration rate with trade credit. J. Math. Stat. **5**(4), 330–333 (2009)
2. R.K. Yadav, A.K. Vats, A deteriorating inventory model for quadratic demand and constant holding cost with partial backlogging and inflation. IOSR J. Math. **10**(3), 47–52 (2014)
3. C. Lin, B. Tan, W.-C. Lee, An EOQ model for deteriorating items with time-varying demand and shortages. Int. J. Syst. Sci. **31**(3), 391–400 (2000)
4. S.P. Aggarwal, C.K. Jaggi, Ordering policies of deteriorating items under permissible delay in payments. J. Oper. Res. Soc. **36**, 658–662 (1995)
5. B.C. Giri, A. Goswami, K.S. Chaudhuri, An EOQ model for deteriorating items with time varying demand and costs. J. Oper. Res. Soc. **47**, 1398–1405 (1996)
6. P. Parvathi, S. Gajalakshmi, A fuzzy inventory model with lot size dependent carrying cost/holding cost. IOSR J. Math. **7**, 106–110 (2013)
7. M. Omar, R. Sarker, W.A.M. Othman, A just-in-time three-level integrated manufacturing system for linearly time-varying demand process. Appl. Math. Model. **37**, 1275–1281 (2013)
8. R. Patel, R.U. Parekh, Inventory model for variable deteriorating items with two warehouses under shortages, time varying holding cost, inflation and permissible delay in payments. Int. Refereed J. Eng. Sci. **3**(8), 06–17 (2014)
9. S.K. Goyal, Economic order quantity under conditions of permissible delay in payments. J. Oper. Res. Soc. **36**, 335–338 (1985)
10. J. Jagadeeswari, P.K. Chenniappan, An order level inventory model for deteriorating items with time â—quadratic demand and partial backlogging. J. Bus. Manag. Sci. **2**(3), 79–82 (2014)
11. R. Venkateswarlu, R. Mohan, An inventory model for time varying deterioration and price dependent quadratic demand with salvage value. Indian J. Comput. Appl. Math. **1**(1), 21–27 (2013)
12. S. Singh, An inventory model for perishable items having constant demand with time dependent holding cost. Math. Stat. **4**(2), 58–61 (2016)
13. S. Khanra, S.K. Ghosh, K.S. Chaudhuri, An EOQ model for a deteriorating item with time dependent quadratic demand under permissible delay in payment. Appl. Math. Comput. **218**, 1–9 (2011)
14. J. Min, Y.-W. Zhou, G.-Q. Liu, S.-D. Wang, An EPQ model for deteriorating items with inventory-level-dependent demand and permissible delay in payments. Int. J. Syst. Sci. **43**(6), 1039–1053 (2012)
15. N.H. Shah, Inventory model for deteriorating items and time value of money for a finite time horizon under the permissible delay in payments. Int. J. Syst. Sci. **37**(1), 9–15 (2006)
16. R. Uthayakumar, S.K. Karuppasamy, A pharmaceutical inventory model for healthcare industries with quadratic demand, linear holding cost and shortages. Int. J. Pure Appl. Math. **106**(8), 73–83 (2016)
17. V.K. Mishra, L.S. Singh, Deteriorating inventory model with time dependent demand and partial backlogging. Appl. Math. Sci. **4**(72), 3611–3619 (2010)

18. S. Khanra, K.S. Chaudhuri, A note on an order level inventory model for a deteriorating item with time dependent quadratic demand. Comput. Oper. Res. **30**, 1901–1916 (2003)
19. J.T. Teng, C.T. Chang, Economic production quantity models for deteriorating items with price- and stock-dependent demand. Comput. Oper. Res. **32**, 297–308 (2005)
20. H.J. Chang, C.H. Hung, C.Y. Dye, An inventory model for a deteriorating items with linear trend in demand under conditions of permissible delay in payments. Prod. Planning Control **12**, 272–282 (2001)

Histogram of Oriented Gradients for Image Mosaicing

Ridhi Arora and Parvinder Singh

Abstract Image Stitching can be defined as a process of combining multiple input images or some of their features into a single image without introducing distortion or loss of information. The aim of stitching is to integrate all the imperative information from multiple source images to create a fused mosaic. Therefore, the newly generated image should contain a full description of the scenery than any of its individual source images and is more suitable for human visual and machine perception. With the recent rapid developments in the domain of imaging technologies, mosaicing has become a reality in wide fields such as remote sensing, medical imaging, machine vision, and the military applications. Image Stitching therefore provides an effective way by reducing huge volume of information by extracting only the useful information from the source images and blending them into a high-resolution image called mosaic.

Keywords Image mosaicing · Image reconstruction from gradient measurements RANSAC estimation · Fundamental matrix · Geometric transformation HoG

1 Introduction

There are many applications where images with larger field of view are of great importance. In areas ranging from medical imaging to computer graphics and satellite imagery, a computationally effective and easy to implement a method to produce high-resolution wide angle images will continue to draw research interest. This production of wide angle image from source images is called Image Stitching. Research into algorithmic image stitching requires image registration, alignment, calibration, and blending which is one of the oldest topics in computer vision. Image stitching for the purposes of mosaic creation can be used by amateur pho-

R. Arora (✉) · P. Singh
Department of CSE, DCRUST, Sonipat (Murthal), Haryana, India
e-mail: ridhiarora584@gmail.com

© Springer Nature Singapore Pte Ltd. 2018   211
B. Panda et al. (eds.), *Innovations in Computational Intelligence*, Studies in
Computational Intelligence 713, https://doi.org/10.1007/978-981-10-4555-4_14

tographers and professionals without requiring detailed knowledge, using digital photography software such as Photoshop that provides easy-to-use instructions and interfaces. Depending on the scene content, the future panorama, and luminance, two major steps are being followed. The first is the registration, that is to be done on the source images, and second is choosing an appropriate blending algorithm.

Registration is the process of spatially aligning images. This is done by choosing one of the images as the reference image and then finding geometric transformations which map other images on to the reference frame. Upon completion of this step, an initial mosaic is created which is made by simply overlapping the source images by the common region in them. This helps in reducing visual and ghosting artifacts. Various registration techniques are described in [1].

The initial mosaic may contain visual and ghosting artifacts because of varying intensities in each of the source images. They can be significantly reduced by choosing an appropriate blending algorithm through which transition from one image to other becomes imperceptible. Good blending algorithms should produce seamless and plausible mosaics in spite of ghosting artifacts, noise, or lightning disturbances in the input images. Previously, proposed blending algorithms work on image intensities or image gradient, at full resolution of the image or at multiple-resolution scales.

Image Stitching is performed by adopting techniques such as feature-based and direct techniques. Direct techniques also called as area-based technique [2] minimize the sum of absolute differences between the overlapping pixels. This technique is computationally ineffective as it takes every pixel window into consideration. The use of rectangular window to find the similar area limits the application of these methods. This method is generally used with applications that have small translation and rotation because they are not invariant to image scale and rotation [3]. The main *advantage* of direct technique is that they incorporate every pixel value which optimally utilizes the information in image alignment but fail to correctly match the overlapping region due to limited range of convergence. On the other hand, feature-based technique works by extracting a sparse set of essential features from all the input images and then matching these to each other [4]. Feature-based methods are used by establishing correspondences between points, lines, edges, corners, or any other shapes. The main characteristics of robust detectors include invariance to scaling, translation, image noise, and rotation transformations. There are many feature detector techniques existing some of which are Harris corner detection [5], Scale-Invariant Feature Transform (SIFT) [6, 7], Speeded Up Robust Feature (SURF) [3, 8], Features from Accelerated Segment Test (FAST) [9], HoG [10], and ORB [11].

The use of image stitching in real-time applications has proved to be a challenging field for image processing. Image stitching has found a variety of applications in microscopy, video conferencing, video matting, fluorography, 3D image reconstruction, texture synthesis, video indexing, super resolution, scene completion, video compression, a satellite imaging, and several medical applications. Stitched images (mosaics) are also used in topographic mapping. For videos, additional challenges are imposed on image stitching. As videos require motion of pictures with varying intensities so, camera feature-based techniques which aim to determine a relationship zoom, and to visualize dynamic events impose between the images is used which poses additional challenges to image stitching.

2 Related Work

Image Stitching is a process of combining source images to form one big wide image called mosaic irrespective of visual and ghosting artifacts, noise addition, blurring difficulties, and intensity differences. For stitching, many techniques have been proposed in many directions and fields. They include gradient field, stitching images in mobiles, SIFT algorithm, SURF algorithm, Haar wavelet method, corner method, and many more.

A well-known intensity-based technique is *feathering* [12], where the mosaic is generated by computing a weighted average of the source images. In the composite mosaic image, pixels are assigned weights proportional to their distance from the seam, resulting in a smoother transition from one image to the other in the final mosaic. But this can introduce blurring, noise, or ghosting effect in the mosaic when the images are not registered properly. Therefore, the intensity-based multiscale method is presented that relies on *pyramid decomposition*. The source images are decomposed into band-pass components, and the blending is done at each scale, in a transition zone inversely proportional to the spatial frequency content in the band.

Gradient field is also exclusively utilized for mosaicing. Sevcenco et al. [13] presented a gradient-based image stitching technique. In this algorithm, the gradients of the two input images are combined to generate a set of gradients for the mosaic image and then to reconstruct the mosaic image from these gradients using the Haar wavelet integration technique.

Further, advancements are made in the field of gradients using Poisson image editing. Perez et al. [14] proposed Poisson image editing to do seamless object insertion. The mathematical tool used in the approach is the Poisson partial differential equation with Dirichlet boundary conditions with the Laplacian of an unknown function over the domain of interest, along with the unknown function values over the boundary of the domain. The actual pixel values for the copied region are computed by solving Poisson equations that locally match the gradients while obeying the fixed Dirichlet condition at the seam boundary. To make this idea more practical and easy to use, Jia et al. [15] proposed a cost function to compute an optimized boundary condition, which can reduce blurring. Zomet et al. [16] proposed an image stitching algorithm by optimizing gradient strength in the overlapping area. Here, two methods of gradient domain are discussed. One optimizes the cost functions, while others find derivative of the stitched image. Both the methods produce good results in the presence of local or global color difference between the two input images.

Brown and Lowe [17] use SIFT algorithm to extract and match features in the source images that are located at scale-space maxima/minima of a difference of

Gaussian function. After successful features are mapped, RANSAC algorithm is applied to remove all the unnecessary outliers but include the necessary inliers that are compatible for Homography between the images. Afterward, bundle adjustment [18] is used to solve for all the camera parameters jointly.

Zhu and Wang [19] proposed an effective method of mosaic creation that includes sufficient iterations of RANSAC to produce faster results. For this, multiple constraints on their midpoint, distances, and slopes are applied on every two candidate pairs to remove incorrectly matched pairs of corners. This helps in making effective panoramas with least no RANSAC iteration. Stitched images (mosaics) are also used in topographic mapping and stenography [20]. Stenography is a technique that is used to hide information in images. Various other operations can also be applied on images being stitched [21–24].

Various other techniques are also introduced that either used area-based or feature-based techniques, but an effective technique is introduced in this paper, viz. HoG (Histogram of Oriented Gradients).

3   Proposed Methodology

The main feature for selection and detection used here is HoG (Histogram of Oriented Gradients) [25]. The method is based on evaluating well-normalized local histograms of image gradient orientations in a dense grid. They not only exaggerate the essential feature points, but in turn make small grids around imperative features and make gradients on features which appear densely in the overlapping region between the source images. This method uses linear as well as nonlinear images (images that differ in angle and camera position).

In this work, we have used HoG with RANSAC, with geometric transformation, with Blob method of feature selection and detection. The RANSAC algorithm uses *fundamental matrix*, which estimates the fundamental matrix from corresponding points in linear (stereo) images. This function can be configured to use all corresponding points or to exclude outliers by using a robust estimation technique such as random sample consensus (RANSAC). Other method makes use of *geometric transformation*. Geometric transform returns a 2D geometric transform object, tform. The tform object maps the inliers in matchedPoints1 to the inliers in matchedPoints2. The matchedPoints1 and matchedPoints2 inputs can be corner point objects, SURF Point objects, MSER objects, or M-by-2 matrices of [x, y] coordinates. The next method is *Blob* method. BLOB stands for Binary Large OBject and refers to a group of connected pixels in a binary image. The term *large* indicates that only objects of a certain size are of interest as compared to those *small* binary objects which are usually considered as noise which in turn is significantly reduced by applying a binary mask that makes the final mosaic independent of ghosting effect and visible seams.

3.1 Main Steps of Image Stitching

All stitching can be divided into several steps. First, registration for the image pair is done. Registration is done on the overlapping region to find the translation which aligns them. It is done to make the images photometrically and geometrically similar. After successful registration, some feature-based methods are applied to extract fine and strongest points from the input images in order to match images to their maximum. Next, these matched points are used either to create Homography using RANSAC or are used in gradient medium to reconstruct the final mosaic.

3.1.1 Feature Matching

Feature matching requires similar features from input images; therefore, it is the integration of the direct method and feature-based method. Direct method matches every pixel of one image with every pixel of other images; i.e., it incorporates every pixel value which makes it less popular technique of feature matching. It is generally used with images having large overlapping region and small translation and rotation because it cannot effectively extract overlapping window region from the referenced images, whereas feature-based method is used over large overlapping region as they only cull requisite and important features. Many feature extraction algorithms have been used such as FAST and Harris corner detection method. FAST features return corner point object points. The object contains information about the feature points detected in a 2D grayscale input image. The detection of FAST feature function uses the features from accelerated segment test (FAST) algorithm to find feature points.

3.1.2 Image Matching

In the following step, the SURFs extracted from the input images are matched against each other to find nearest-neighbor for each feature. Connections between features are denoted by green lines. When using RANSAC, unnecessary lines of matching are removed as outliers and necessary lines remained as inliers. But when geometric transformation is used with HoG to match the images, RANSAC estimation is eliminated because geotransform is a complete package of inlier and outlier detection and removal of unwanted noise from the input set of images.

3.1.3 Image Calibration

When HoG is used with RANSAC for matching images, calibration is done after feature extraction and matching because it returns projective transformations for rectifying stereo images. Calibration rectification function does not require either intrinsic or extrinsic camera parameters. The input points can be M-by-2 matrices of M number of [x, y] coordinates, or SURF Point, MSERs, or corner point object.

4 Experimental Setup

4.1 Performance Matrices

The *Peak Signal-to-Noise Ratio (PSNR)* and *mean squared error (MSE)* are the two error metrics used to compare mosaic quality. The MSE represents the mean squared error between the final (de-noised image) and the original image, whereas PSNR represents a measure of the peak error.

4.1.1 PSNR

As the name suggests, peak signal-to-noise ratio is the measure of peak error which is used to depict the ratio of maximum possible power of signal (image) to the power of the corrupting noise that affects the fidelity of its representation. It is represented in terms of mean square error (MSE) as follows:

$$PSNR = 10 * \log 10(256^2/mse)$$

PSNR being popular and accurate in terms of prediction is commonly used in image processing. The higher the PSNR value, the better the quality of the reconstructed final mosaic.

4.1.2 MSE

Mean square error is the term used to present the average between the estimator and what is estimated, i.e., between the final (de-noised image) and the original image before introduction of noise.

$$MSE = sum1/numel \ (image)$$

This enables us to compare mathematically as to which method provides better results under same conditions like image size noise.

5 Experimental Results

In this section, the performance of the proposed method is illustrated with experiments. Given two partially overlapping, previously registered images with photometric inconsistencies (i.e., with differences in light intensity within the overlap region), the objective is to stitch them and produce a larger mosaic image which looks smooth and continuous without any noticeable artifacts such as seams or blurring.

The two images have an overlapping region indicated by a vertical black patch. The intensity levels in the two images have been modified and are clearly different. This has the effect in a "direct paste" mosaic image. The result of pasting the two images directly is shown in Fig. 1c. Later, when the HoG method (in Fig. 2a, b) is applied, the resultant seam is seen to have been absorbed and a final segmented panorama is produced in Fig. 3.

Later, the mosaic is improved upon by removing noise and ghosting artifacts from the generated panorama which can be visualized in Fig. 4.

(a)   **(b)**   **(c)**

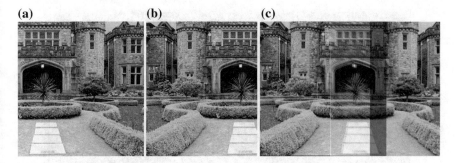

Fig. 1 From *left* to *right,* **a** and **b** are original images, and **c** is the result of direct pasting

(a)   **(b)**

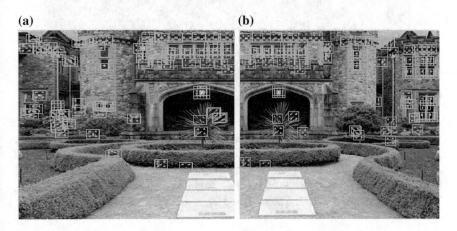

Fig. 2 From *left* to *right,* **a** and **b** shows the HoG descriptor in the input images

Fig. 3 Final segmented panorama

Fig. 4 Output mosaic without ghosting

The above experiments illustrate that the proposed method can lead to mosaic images without any visual artifacts despite differences in the intensities of the input images in the overlapping region. The algorithm was implemented using MATLAB and was running on a ×86 64 bit PC Architecture, Intel Core Duo T2080, 1.73 GHz (Figs. 5, 6 and 7; Tables 1 and 2).

Fig. 5 Temporal comparison

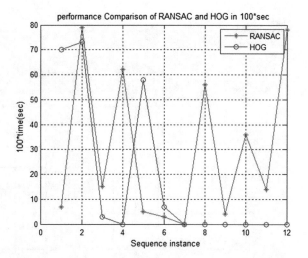

Fig. 6 In search for strongest points in both input images, the matched points are compared against each other

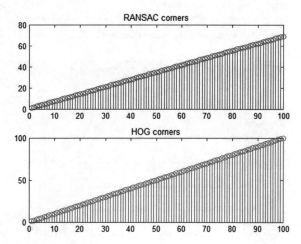

p1,p2 = image inputs; Im1 = image output;
X = imread ('cameraman.tif');
X1 = X;
X1(X <=100) = 1;
[psnr,mse,maxerr,L2rat] = measerr (X, X1)
To find energy distribution in image:
Entropy (p1)
Entropy (p2)
Entropy (Im)
PSNR—Peak (pixel) SNR (psnr = 10*log10 (256^2/mse));
MSE—Mean Square Error (mse = sum1/numel (image));

(a)

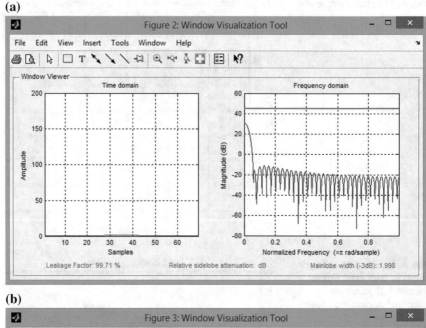

(b)

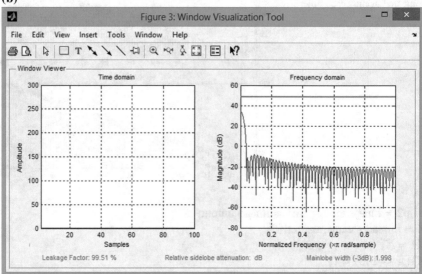

Fig. 7 a and **b** depicts the inliers comparison in time and frequency domain of HoG and RANSAC

Ghosting: If objects are not static in the overlap image, the image will appear blurry and ghosted. So far, scholars both domestic and foreign have done a lot of researches and made remarkable achievements in de-ghosting. Shum and Sezliski [26] also proposed a method for eliminating small ghosting based on computing optic flow and then doing a multiway morph.

Table 1 Image properties comparison

	PSNR (p1)	PSNR (p2)	PSNR (lm1)	MSE (p1)	MSE (p2)	MSE absolute (lm1)	Maxerr (p1 ∼ p2 ∼ lm1)	L2RAT (p1)	L2RAT (p2)	L2RAT (lm1)
Blob_Ransac 200a500.m	18.6938	19.8816	20.3460	878.4119	668.2186	600.4480	99	0.9602	0.9724	0.9687
Blob_Ransac 1000a10000.m	18.6938	19.8816	20.3569	878.4119	668.2186	598.9536	99	0.9602	0.9724	0.9688
Blob_Ransac 2k10k.m	18.6938	19.8816	20.3246	878.4119	668.2186	603.4152	99	0.9602	0.9724	0.9689
Hog_geo 1000a10000.m	18.6938	19.8816	20.3102	878.4119	668.2186	605.4297	99	0.9602	0.9724	0.9686
Hog_geo 200a2k.m	18.6938	19.8816	20.2615	878.4119	668.2186	612.2477	99	0.9602	0.9724	0.9683
Hog_geo 100a1000.m	18.6938	19.8816	20.2649	878.4119	668.2186	611.7767	99	0.9602	0.9724	0.9681
Hog_Ransac 1000a20000.m	18.6938	19.8816	20.3772	878.4119	668.2186	596.1544	99	0.9602	0.9724	0.9688
Hog_Ransac 600a5000.m	18.6938	19.8816	20.3772	878.4119	668.2186	596.1544	99	0.9602	0.9724	0.9688
Hog_Ransac 300a10000.m	18.6938	19.8816	20.4004	878.4119	668.2186	592.4782	99	0.9602	0.9724	0.9688

Table 2 Performance comparison of methods used

Features	Unitary mode detection—Blob linear spatial RANSAC	Gradient HoG linear spatial RANSAC	Gradient HoG nonlinear spatial geometric transformation
Detection	Only one detection mode—SURF	Two detection methods are used: FAST and HoG	Two detection methods are used: FAST and HoG
Algorithms and no. of corner points	'Metric Threshold' = 2000 and compare strongest feature points	FAST detection for 300 level points, with strong feature point comparison	FAST detection for 300 level points, with strong feature point comparison
Feature detection iterations	First 2000 followed by 10,000 RANSAC trials	First 300 followed by 10,000 RANSAC trials	First 200 followed by 2000 trials
Spatial reference	Linear image spatial—no reference to light source/viewpoint/field-of-depth/background reference	With reference to light source/viewpoint/field-of-depth/background reference	With reference to light source/viewpoint/field-of-depth/background reference
Directional spatial reference	No directional reference	Gradient type detection	Gradient type detection
Blending	Via Alpha Blender	Via Alpha Blender	Via Alpha Blender

But here, an impression-based method HoG is used to get better PSNR and MSE values. The proposed algorithm shows better results than the standard median filter (MF) and decision-based algorithm (DBA). The method performs well in removing low-to-medium density impulse noise with detailed preservation up to a noise density of 70%, and it gives better peak signal-to-noise ratio (PSNR) and mean square error (MSE) values.

6 Conclusion

Image gradient blending techniques prove effective because the human visual system is known to be more sensitive to local contrast changes than to global intensity variations. Therefore, the gradient domain provides an excellent framework for image processing applications such as image editing, high dynamic range imaging, and compression and image mosaicing. In all these methods, the gradients of the source images are modified and the modified gradients are used to obtain the final image. In general, these gradient modifications lead to a gradient which typically is not a conservative vector field, and image reconstruction from this gradient no longer has an exact solution, but can be formulated as an optimization problem.

Therefore, a method of seamless stitching of images with photometric and geometric inconsistencies in the overlapping region has been discussed in this dissertation. This is done by generating a set of stitched gradients of the input images and then reconstructing the mosaic from the stitched gradients. This requires solving Poisson equation for the input images leading to mosaic formation without visual artifacts. Experimental results illustrate the method and show that it can lead to seamless mosaic images despite intensity differences in the overlap region of the input images.

7 Future Scope

The given work was based on the implementation and evaluation of the gradient-based HoG mosaicing technique. An approach and methodology have been proposed to enhance the performance producing the best quality panoramas by the fusion of the complimentary features specific to these algorithms. The test input images used for the present work were the planar images, but it can further be extended for MR images and cylindrical or spherical images.

Moreover, for computational enhancements, optimization of the algorithms used can be performed to dramatically increase timing results. Currently, the code developed for the proposed system runs mostly in MATLAB and is not designed especially for a high speed of computation. Multithreaded software or GPU-based algorithm implementations are all ways to reduce the time of computation, making use of parallelism in different stages of the mosaicing process. These techniques can be implemented in addition to actual optimization of the mosaicing algorithms and procedures themselves.

References

1. L.G. Brown, A survey of image registration techniques. Comput. Surv. **24**(4), 325–376 (1992)
2. B.Z.B. Zhang, Computer vision vs. human vision, Cogn. Informatics (ICCI), in *2010 9th IEEE International Conference* (2004), p. 318108
3. Survey (ebtsam, elmogy)
4. CV_Overview
5. C. Guo, X. Li, L. Zhong, X. Luo, A fast and accurate corner detector based on Harris algorithm, in *3rd International Symposium Intelligence Information Technology Applied IITA 2009*, vol. 2, no. 4, pp. 49–52 (2009)
6. L. Zeng, S. Zhang, J. Zhang, Y. Zhang, Dynamic image mosaic via SIFT and dynamic programming. Mach. Vis. Appl. **25**(5), 1271–1282 (2014)
7. J. Melorose, R. Perroy, S. Careas, No Title No Title. Statew. Agric. L. Use Baseline **1**, 2015 (2015)
8. Z. Wang, F. Yan, Y. Zheng, An adaptive uniform distribution surf for image stitching, in *2013 6th International Congress Image Signal Processes*, vol. 2, (2013), no. Cisp, pp. 888–892
9. E. Rosten, T. Drummond, Machine learning for high speed corner detection, *Computer Vision —ECCV 2006*, (2004), pp. 430–443
10. N. Dalal, W. Triggs, Histograms of oriented gradients for human detection, in *2005 IEEE Computer. Social Conferences Computer Vision Pattern Recognition CVPR05*, vol. 1, (2004), no. 3, pp. 886–893
11. A. Kulkarni, J. Jagtap, V. Harpale, *Object recognition with ORB and its implementation on FPGA*, vol. 3 (The accents Organization, 2013)
12. R. Szeliski, Video mosaics for virtual environments. IEEE Comput. Graph. Appl. **16**(2), 22–30 (1996)
13. I.S. Sevcenco, P.J. Hampton, P. Agathoklis, Seamless stitching of images based on a HAAR wavelet 2D integration method. Department of Electrical and Computer Engineering University of Victoria (2011)
14. P. Pérez, M. Gangnet, A. Blake, Poisson image editing. ACM Trans. Graph. **22**(3), 313 (2003)
15. J. Jia, J. Sun, C.-K. Tang, H.-Y. Shum, Drag-and-drop pasting. Siggraph **25**(3), 631 (2006)
16. A. Levin, A. Zomet, S. Peleg, Y. Weiss, Seamless image stitching in the gradient domain. Eccv **3024**, 377–389 (2004)
17. M. Brown, D.G. Lowe, Recognising panoramas, in *Proceeding IEEE International Conference Computer Vision*, (2003) pp. 1218–1225
18. B. Triggs, P. Mclauchlan, R. Hartley, A. Fitzgibbon, Bundle adjustment—a modern synthesis 1 introduction. Vis. Algorithms Theory Pract. **34099**, 1–71 (2000)
19. M. Zhu, W. Wang, B. Liu, J. Huang, A fast image stitching algorithm via multiple-constraint corner matching. Math. Prob. Eng. **2013** (2013)
20. P. Singh, S. Batra, H.R. Sharma, Steganographic methods based on digital logic, in *Proceedings of the 6th WSEAS International Conference on Signal Processing, Dallas, Texas, USA* (2007), pp. 157–162

21. A.V. Krishna, Performance evaluation of new encryption algorithms with emphasis on probabilistic encryption and time stamp in network security, vol. 3, no. 5, pp. 39–46 (2009)
22. D. Hooda, P. Singh, A comprehensive survey of video encryption algorithms. IJCA **59**(1), 14–19 (2012)
23. R. Singh, P. Singh, M. Duhan, An effective implementation of security based algorithmic approach in mobile adhoc networks. *Hum—centric Comput. Inf. Sci.* **4**(1), 7:1–7:14 (2014)
24. D. Singh, D. Sethi, Reduction of noise from speech signal using HAAR and biorthogonal wavelet. **7109**, pp. 263–269 (2011)
25. G. Tsai, Histogram of oriented gradients. Lecture Series University of Michigan (2010), pp. 1–6
26. H.-Y. Shum, R. Szeliski, Construction of panoramic mosaics with global and local alignment. Int. J. Comput. Vis. **36**(2), 101–130 (2000)

Capacitive Coupled Truncated Corner Microstrip Stack Patch Antenna

Suraj R. Balaskar and Neha Rai

Abstract A design of a circular polarized microstrip stack patch antenna is presented. In this paper, a corner truncation at opposite side of square patch is printed on two FR-4 substrates, one is for stack and another for radiating patch, to achieve circular polarization; the rectangular slot is introduced at the centre in diagonal with axis to increase bandwidth and reduce size. The design patch is suspended from ground plane, and air gap is introduced between it. To reduce the inductive effect from long feeding probe to radiating patch a capacitive feed is placed, series opposition of inductive and capacitive will balance input impedance. To increase the bandwidth and reduce the size of the designed antenna, a shorting post is introduced to obtain maximum current distribution along the patch. The proposed design combines the effective features such as large bandwidths, low volume, simple design and robust structure, and the patch is simulated in IE3D software operating at 5.2 GHz band. Simulated results show that the antenna achieves 10-dB return loss bandwidth of 52.29% (3.69–6.20 GHz) and 3-dB axial ratio bandwidth is of 17.69% (4.92–5.84 GHz) with the antenna gain level at about 3.68 dBi.

Keywords Axial ratio · Capacitive coupling · Circular polarization
Corner truncated · Rectangular slot · Shorting pin · Stack patch

S.R. Balaskar (✉) · N. Rai
Electronics and Telecommunication, Pillai HOC College of Engineering and Technology,
Mumbai University, Rasayani, Raigad 410201, Maharashtra, India
e-mail: suraj.balaskar@rediffmail.com

N. Rai
e-mail: nrai@mes.ac.in

© Springer Nature Singapore Pte Ltd. 2018
B. Panda et al. (eds.), *Innovations in Computational Intelligence*, Studies in
Computational Intelligence 713, https://doi.org/10.1007/978-981-10-4555-4_15

227

1 Introduction

The microstrip patch antenna is a very important device nowadays, because it has several advantages than other patch antenna such as low volume, easy to conformal with compact devices, provide both linear and circular polarization, easy to manufacture and low fabrication cost.

Today, most of the applications need large bandwidth and low volume, so that they can be easily interfaced with compact low power consumption devices and also they provide high degree of isolation between the different types of polarization. The circularly polarized microstrip antennas are mostly used in satellite communication, mobile communication and radar because of target independency or device orientation. As started in 1970, many authors have been working on how to increase bandwidth and reduce size of antenna and these two factors limit its practical application [1]. There are many broadband and compact techniques to overcome its technical challenges such as placing the shorting post between ground and radiating patch, introducing the rectangular slot and stack patch to increase the bandwidth, to achieve the circular polarization a truncated corner has used, to achieve the large impedance bandwidth a capacitive coupling probe feed methods has used to compensate inductive reactance offered by long feeding probe. The designed patch is suspended on air and dielectric to increase the height of antenna and also compensate the increasing dielectric constant. The stack and radiating patches have same geometry but different dimension. The simulated result indicates that its return loss bandwidth or s11 bandwidth is 48%. Pan et al. [2] presented the idea of introducing the single feed patch by shorting pin. The given antennas indicate the large axial ratio bandwidth. In this paper, we have presented single coaxial feed capacitive coupled corner truncated stack patch with rectangular slot on radiating and stack patches both placed on two FR-4 substrate, air gap between ground and radiating patch is introduced and shorting pin is placed between ground conducting plane and radiating patch.

2 Proposed System

2.1 Top View

2.1.1 Stack Patch

The proposed antenna consists of corner truncated square patch length $L_s = 10.10$ mm and width is $W_s = 10.10$ mm and truncation level $S_s = 5$ mm on opposite side of patch placed on top of FR-4 substrate thickness is about $h = 1.6$ mm, height and width of FR-4 is $L_g = 40$ mm ($\varepsilon_r = 4.4$, tan $\delta = 0.02$) at a height of $g_s = 14.8$ mm on ground plane. The rectangular slot is introduced at the centre of axis at diagonal orientation on stack patch with the dimension $L_s = 2.825$ mm and $W_s = 1$ mm as shown in Fig. 1a.

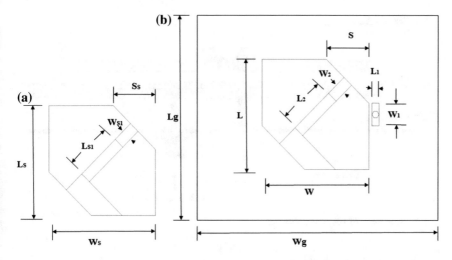

Fig. 1 Top view of proposed antenna **a** stack patch, **b** radiating patch

2.1.2 Radiating Patch

The radiating patch consists of corner truncated square patch length $L = 10$ mm and width is $W = 10$ mm and truncation level $S = 5$ mm on opposite side of patch placed on top of FR-4 substrate thickness is about $h = 1.6$ mm, height and width of FR-4 is $L_g = 40$ mm ($\varepsilon_r = 4.4$, tan $\delta = 0.02$) at a height of $g_r = 7.4$ mm on ground plane. The rectangular slot is introduced at the centre of axis at diagonal orientation on radiating patch with the dimension $L_2 = 5.6$ mm and $W_2 = 1.6$ mm as shown in Fig. 1b.

2.2 Side View

Initially, a ground plane is considered as infinite for simulation with height $W_g = 40$ mm and width is $L_g = 40$ mm placed at $Z = 0$, SMA connector with inner conductor extended up to $z = 7.4$ mm to radiating patch placed at $x = 10.12$ mm with inner conductor extended up to $z = 7.4$ mm from ground plane is attached to rectangular patch $L1 = 1.3$ mm, $W1 = 3.6$ mm capacitive feeding is observed between rectangular and main driven patch to reduce the inductive reactance offered by length of coaxial inner conductor, since series opposition of inductor (length of coaxial inner conductor) and capacitor (distance between radiating and feed patch) will cancel each other. A shorting pin at $z = 0$ (ground plane) and $Z = 7.4$ (radiating patch) at $x = -2.5$ and $y = 0$ is placed. Figure 2 shows the details of the structure of proposed antenna where $g = 5.8$ mm as shown in Fig. 2.

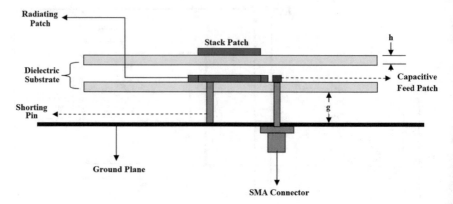

Fig. 2 Side view of proposed antenna

3 Parametric Studies

3.1 Performance of Antenna on Design Parameters

A. Effect of air gap (g)

The important parameter in these proposed antennas is air gap placed between the ground plane, radiating patch and stack patch. This acts as a dielectric medium for microstrip patch antennas, and it decides the bandwidth of antennas. As the effective height of the substrate increases or dielectric permittivity decreases (inverse relation between permittivity and height of patch), it results into wide bandwidth [3]. The parallel relation between quality factor of ground, radiating patch and dielectric materials is added. If we introduced another patch above the proposed structure, the quality factor goes on decrease and consequently bandwidth goes on increase (inverse relation between BW and QF). The effect of air gap on impedance bandwidth is also affected by dielectric materials, distance of radiating patch and feeding patch, probe diameter and the dimensions of feeding patch and radiating patch. BW enhancement can be done by properly adjusting the above parameters [4, 5].

B. Effect of distance between feeding and radiating patch (d)

The small variation of dimensions of rectangular feeding patch length $L1$ and width $W1$ along with location of feed points has strong effect on wide impedance bandwidth [6]. The input impedance of an antenna is reactive because of the extended inner conductor of probe feed and resistive parts depend upon the separation distance. If the distance between radiating patch and feed patch increases the resistive part decreases and inductive reactive increases consequently. VSWR bandwidth is increased. Thus, by changing the separation distance between feeding and radiating patches, the impedance bandwidth is also changed [4, 5]. Figure 3 shows the

Fig. 3 Effect of parameter variation on the input impedance

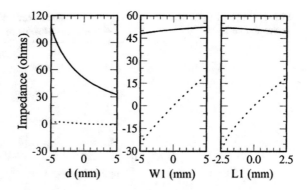

dimension of feed strip and separation distance versus input impedance of proposed antenna. It is clearly observed that width and length of the feed strip have very significant impact on input impedance. But separation distance provides linear relation.

C. Effect of shape of the radiating patch

Impedance bandwidth of microstrip patch antenna can be increases by 50% [4], if the regular shape radiating patch with probe feed capacitive coupled is used. Further, 10–15% bandwidth can be increased by incorporating rectangular, square or triangular slot both on radiating and nonradiating patches at the centre to enhance the fringing field and bandwidth. This rectangular slot increases the large area for fringing field at the edge of the patch so voltage distribution is increased along the patch. Also the inductive reactance of designed patch is balanced and large impedance bandwidth is achieved [4, 5].

4 Experimental Results

The designed patch is simulated on IE3D software. The resonant frequency of 5.2 GHz is chosen. As shown in the simulated result, Fig. 4 shows the graph of axial ratio versus frequency of circular polarized antenna, which achieves the 3-dB AR bandwidth of about 17.69% (4.92–5.84 GHz).

As shown in the simulated result, Fig. 5 shows the graph of total field gain versus frequency of proposed antenna. The achieved result is 3.68 dBi at 5.2 GHz.

Figure 6 shows the simulated result of S11 versus frequency of the proposed antennas. The achieved result is 10-dB return loss bandwidth of about 52.29% (3.69–6.20 GHz).

Figure 7 shows the current distribution of the proposed antennas with shorting post. Due to the use of shorting post between grounds and radiating patch, the current distribution among the patch is increased as compared to Fig. 8.

Fig. 4 Graph of axial ratio
versus frequency

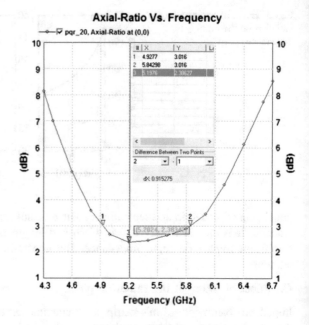

Fig. 5 Graph of total field
gain versus frequency

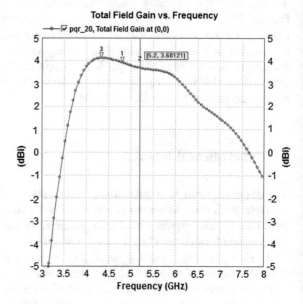

Figure 8 shows the current distribution of the proposed antennas without shorting post. As shown in the figure, the current distribution of radiating patch is not uniform along the patch (Figs. 9 and 10).

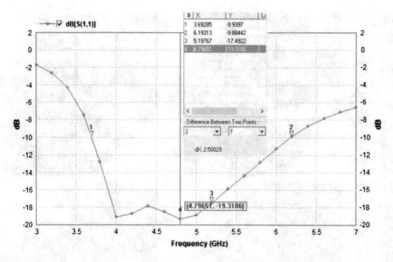

Fig. 6 Graph of S11 versus frequency

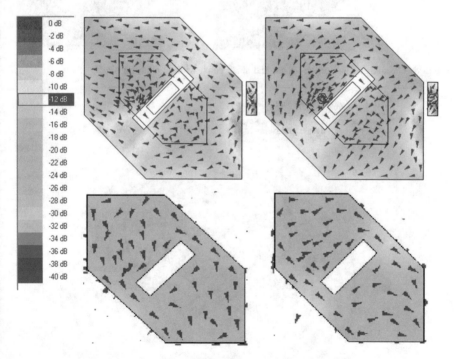

Fig. 7 Current distribution on driven and stack patch with shorting pin

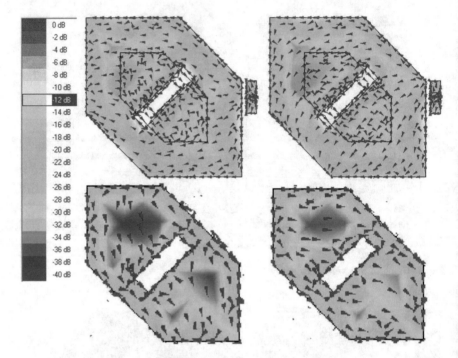

Fig. 8 Current distribution on driven and stack patch without shorting pin

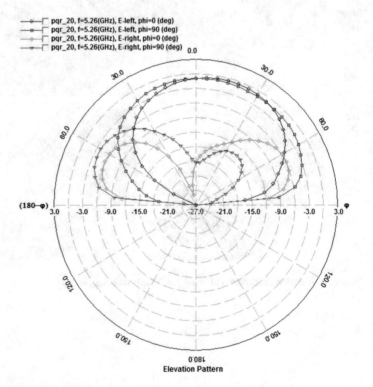

Fig. 9 2D radiation pattern

Fig. 10 Photograph of
proposed antenna (*top view*
and *side view*)

The simulation shows good agreement with the experimental results of the
proposed antenna in Fig. 11. The 10-dB impedance bandwidth of the measured
return loss is approximately similar to the simulated result.

5   Conclusion

In this paper, a technique to design a circular polarized truncated corner microstrip
stack patch antennas with shorting pins to increased bandwidth and reduces the
height of antennas, rectangular slot on radiating patch to increase fringing effect and
truncated corner to achieve the circular polarization is presented. A prototype
structure is designed and operating at 5.2 GHz is simulated in IE3D software and
measured to validate the concept for practical application. The proposed systems
shows the 10-dB return loss bandwidth is 52.29% (3.69–6.20 GHz), 3-dB AR
bandwidth of about 17.69% (4.92–5.84 GHz) and a peak gain of 3.68 dBi is

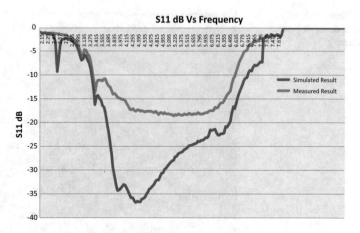

Fig. 11 Measured and simulated result of proposed antenna

achieved. The measured and simulated result is shown in Fig. 11. Due to the use of rectangular slot on main driven patch and stack patch structure on air, the bandwidth is improved. The designed patch has a low volume, good radiation over the given band, large BW; simple structure balanced input impedance as well as easiness of design which makes it a good choice for broadband CP applications.

References

1. G. Kumar, K.P. Ray, *Broadband Microstrip Antennas* (Chap. 8) (Artech House, Norwood, MA, USA, 2003)
2. S.C. Pan, K.-L. Wong, *Design of Dual Frequency Microstrip Antennas Using a Shorting Pin Loading*. 7803–4478-2/98/1998 (IEEE, 1998), pp. 312–315
3. C.A. Balanis, *Antenna Theory Analysis and Design*, 2nd edn. (Wiley, Hoboken, NJ, USA, 2004)
4. A. Asthana, B.R. Vishvakarma, Analysis of gap coupled microstrip antenna. Int. J. Electron. **88**(6), 707–718 (2001)
5. V.G. Kasabegoudar, D.S. Upadhyay, Design studies of ultra-wideband microstrip antennas with a small capacitive feed. Int. J. Antennas Propag. **67503**. doi:10.1155/2007/6750 (2007)
6. G. Mayhew-Ridgers, J.W. Odendaal, J. Joubert, Single-layer capacitive feed for wideband probe-fed microstrip antenna elements. IEEE Trans. Antennas Propag. **51**(6), 1405–1407 (2003)
7. M.J. Alexander, Capacitive matching of microstrip patch antennas. IEEE Proc. **136**, 172–174 (Pt. H, No. 2) (1989)
8. Y. Liu, J. Liu, P. Li, in *Designing a Novel Broadband Microstrip Antenna with Capacitive Feed*. 2011 Cross Strait Quad-Regional Radio Science and Wireless Technology Conference, 978-1-4244-9793-5/11/2011 (IEEE, 2011), pp. 156–159
9. Z.B. Wang, S.J. Fang, S.Q. Fu, S.L. Jia, Single-fed broadband circularly polarized stacked patch antenna with horizontally meandered strip for universal UHF RFID applications. IEEE Trans. Microw. Theor. Technol. **59**(4), 1066–1073 (2011)

10. R.A. Sainati, *CAD of Microstrip Antennas for Wireless Application* (Chapter 5.2 Single-feed circularly, Polarized Element-114, Chapter 6.5 Capacitively Fed Microstrip Patches), p. 143
11. Y. Li, X. Dai, C. Zhu, in *A Compact and Low-profile Antenna with Stacked Shorted Patch Based on LTCC Technology*. IEEE Conference on Antenna and Propagation Letter, vol. 1, pp. 160–163 (2013)
12. S.I. Al-Mously, A.Z. Abdalla, M.M. Abousetta, *Design of a Broadband Stacked Rectangular MPA with Shorting Pins for GSM-Family and Other Cellular Applications*. 1-4244-1468-7/07/2007/IEEE, pp. 419–422 (2007)

URL-Based Relevance-Ranking Approach to Facilitate Domain-Specific Crawling and Searching

Sonali Gupta and Komal Kumar Bhatia

Abstract The WWW is a vast repository of all the types of information known to mankind and thus is capable of serving the frequent varying needs of its users. Classifying and organizing the webpages according to their domain or topic will help the search engine in retrieving and returning a set of fairly relevant pages to the users. This classification is generally done on the basis of their underlying text or content. This paper brings in a novel approach that tries to predict the relevance of a webpage in a domain not by downloading its content but based on the web documents it is linked to. The approach offers advantages of efficiency in cost and performance as the most easily and the least expensive information available about a webpage is its uniform resource locator (URL) [1]. Since the URLs serve as the unique identifier, they are assumed to be an important source for the content of a web page, and therefore, the proposed approach associates the domain information with the web pages based on their URLs.

Keywords URL · Domain identification · Topic-specific
Web page classification · Crawler · Search engine · Domain-specific
Focused crawler · Hidden-web crawler

1 Introduction

The activity of searching and browsing the information is typically done by the user with the help of search engines like "Yahoo" and "Google" [2]. It is very important for these search engines to retrieve and return a set of fairly relevant webpages to the user, as rapidly as possible [3, 4]. But, the results provided by the search engines rely on the collection of documents downloaded by the employed crawler. However, it is often certain that the user is only interested in getting the web

S. Gupta (✉) · K.K. Bhatia
Department of Computer Engineering, YMCA University of Science & Technology,
Faridabad, India
e-mail: sonali.goyal@yahoo.com

© Springer Nature Singapore Pte Ltd. 2018   239
B. Panda et al. (eds.), *Innovations in Computational Intelligence*, Studies in
Computational Intelligence 713, https://doi.org/10.1007/978-981-10-4555-4_16

documents on specific domains such as Books and Travel. Accordingly, the search engine must provide domain-specific result pages oriented to the user's domain of preference for which the underlying crawler of the search engine must gather only those pages that seem relevant to that domain. Hence, the crawler used must be made intelligent enough to decide upon the relevance of the web pages as it traverses during the process.

This paper presents an approach that helps the crawler to target only those URLs for downloading that seems relevant to a domain. If the crawler employs some scheme to predict the relevant pages in a domain, it can significantly improve its performance through as follows:

1. The improved quality of its downloaded collection
2. The gain in the rate of gathering the web pages

To predict the relevance of a web page in a domain, its content must be fetched or downloaded by the crawler [3–5]. But, predicting the relevant domain of each and every hypertext document on the Web will exhaust the crawler's resources such as time and bandwidth. Besides this, the task will become infeasible over time due to the ever-expanding nature of the WWW. Thus, a technique that automatically identifies the relevant domain of the web document without fetching its content seems truly important. As the domain of the web page has to be predicted without loading its content or purely on the basis of its associated URL, the problem seems impractical and raises a real challenge [6, 7].

This paper brings in a novel approach toward the idea of predicting the domain of any hypertext document without downloading its content. Such a technique to predict the relevant domain of the webpage without it being downloaded will prove a manifold advantageous to the current-day search engines and further augment their capability of presenting the relevant results to the user. The proposed approach offers the following advantages:

1. Efficient crawling process: The crawler can find the relevant web pages by sifting just through the URLs. If a URL does not seem specific to the target domain of the crawler, it can simply avoid visiting the web page associated to that hyperlink [8].
2. Reduced bandwidth consumption: Avoiding the download of unnecessary web pages saves network bandwidth which can be used for downloading other useful web pages [9].
3. Better search process and organized indexes: The web pages from different domains can be easily downloaded and organized in separate domain-specific indexes or repositories for the purposes of presenting the search results. As the index is organized according to the domain information, it significantly reduces the response time for each user.

As, the only information available for any given web page before it is actually downloaded by the user is only its URL. Thus, the proposed approach targets to associate some domain information with the URLs. This leads to significant savings on user time and network resources like bandwidth etc. Alternatively, the approach targets to find the domain of any URL without downloading its content. The next section presents the framework of the proposed approach as applied to crawling.

2 Proposed Work

A traditional crawler works by taking a URL from the URL frontier, downloading the web page associated with that URL, extracting hyperlinks that are embedded in the downloaded web page, and adding extracted hyperlinks to the URL frontier to keep the process going. It scans the web structure by following the hyperlinks embedded in each page and downloading the associated content.

In the proposed framework, the crawler is initialized by choosing a seed set of URLs in each domain (as shown in Fig.1). These seeds have been stored in the URL pools respective to its domain. For example, the URLs like www.makemytrip. com, https://placetoseeindelhi.wordpress.com which provide information in Travel domain are included in the URL pool meant to store the URLs for Travel domain. Each such group of URLs that belong to a common domain is therefore referred to as domain-specific URL pool.

In other words, the proposed crawler follows a domain-specific approach to initialize the seed URLs for crawling where each URL pool stores the URLs from different domains such as Books, Travel, Auto, Real Estate, Food. This helps the proposed crawler to work well in the domain-specific mode as per requirement.

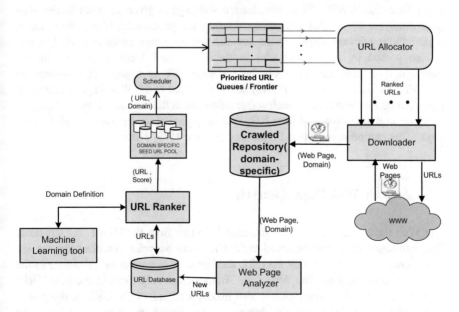

Fig. 1 The proposed architecture of URL-based domain-specific crawler

2.1 Step 1: Initialization

The proposed crawler gathers the seed URLs for the domain-specific URL pools by using two techniques. To start initially, it takes advantage of the classification hierarchy offered by DMOZ as it includes a collection of Web sites and their links that are manually selected and classified by Open Directory volunteer editors. For each category in the hierarchy, the retrieval of top N relevant URLs is supported by the system. These top N URLs will serve as starting points for the crawler. During the crawl, certain URLS will be gathered. All these gathered URLs are added to domain-specific URL pools according to their relevant domain with the help of other components of the crawler.

Also, once each of these domain-specific URL pools is initialized with some URLs, the URLs in these pools need to be prioritized so as to enable the crawler to always populate its document collection with some finite number of top relevant pages in that domain [5, 6, 9]. So, the URL Ranker ranks each URL before adding it to the respective URL pool.

2.2 Step 2: Creating the Crawled Repository

In order to create the downloaded collection for the crawler, the URL Allocator takes the URLs from each of the domain-specific prioritized URL queues and allocates them to the Downloader that establishes HTTP connections to download pages from the WWW. Each downloaded webpage is given as input to the Web page Analyzer for further processing. During the processing, if any information regarding the relevant domain of the downloaded Web page becomes available, that is also passed by the Downloader to the Web Page Analyzer along with the webpage. For example, suppose, the webpage WP is a page that belongs to TRAVEL domain, then (WP, Travel) pair is provided to the Web Page Analyzer. It may also be the case that no such information about the domain of the webpage is acquired in due course, and then NIL is simply passed for the domain of the web page. This can be represented by passing the pair (WP, NIL).

2.3 Step 3: Web Page Analysis

The Web Page Analyzer extracts the useful content from the HTML tag structure. The extracted URLs must be added to the URL pool but prior to that a score termed as **R-Score** is assigned to it by the URL Ranker so as to predict the relevance of the URL in a domain. Thus, the Web Page Analyzer not only extracts the new URLs from the downloaded webpage but also makes a record of the URL of the parent pages of the new URL and the domain of the parent pages of that new URL.

Fig. 2 Example of a web page having new or child URLs

It stores this information in the URL database. So, during the run of the crawler, the URL database is used to store the information as follows:

New/child URL	Parent URL	Domain of parent URL

Consider, for example, the downloaded web page in Fig. 2. The parent and child URLs are represented by the ovals marked in red.

The information that is stored in the URL database corresponding to this downloaded web page and the new URL is represented by the following data structure in the URL database.

New/child URL	Parent URL	Domain of the parent URL
http://www.theweekendleader.com/Travel/1913/must-in-delhi.html	http://placetoseeindelhi.wordpress.com/2014/06/24/delhi-travel-to-capital-city-of-india/	Travel

On the basis of the information that is stored in the URL database, the domain for each URL has to be predicted which is done on the basis of a relevance score (R-Score) that will be computed by the URL Ranker for each URL in each domain of consideration. The process of computing the R-Score is explained in detail in the next section.

2.4 Step 4: R-Score Computation

The URL Ranker is one of the most important components of the proposed crawler. It provides a rank to each URL that is downloaded by the Downloader. This ranking is based on the relevance of the URL in a domain and is important to help the Downloader so as to download more important and better pages earlier during its execution.

The URL Ranker basically takes the URLs from the URL database and ranks them according to their relevance. Ranking the URLs is important so that relevant URLs can be considered on priority basis than the other documents.

Each URL is ranked by the URL Ranker to predict its relevance and importance:

1. Among all the different URLs of a specific domain in the domain-specific URL pool.
2. And across the URLs of all domains in different domain-specific URL pools.

Prior to calculating the relevance score of any URL, the URL Ranker extracts the absolute hyperlinks from their relative counterparts. This becomes needful as the same URL might have been referred in many different documents. Typically, the target of a link relative to the web page WP is defined in the HTML encoding of the link from Web page WP. So, if there is a relative link encoded in the HTML of the page www.easemytrip.com then an absolute URL for it is generated. For example:

<a href="/europe">Explore Europe</a>

points to the URL http://www.easemytrip.com/Europe.

The URL Ranker fetches the URLs from the URL database and computes their relevance score so that URLs can be prioritized for further processing. This relevance score of a URL in domain D has been represented by R-Score (URL, D).

More often, the relevance of any new URL in a domain is not known, thus the relevance score of the new URL has been computed by the URL Ranker before fetching the web page available at that URL. Computing the R-Score has been simplified by considering the hyperlinked structure of the web where similar kinds of pages are linked together by providing hyperlinks.

For computing the R-Score for a new URL, the URL Ranker takes into account the relevance score of its parent page. The parent page is defined as the page that contains the hyperlink embedded for the new URL. Also, in that case the new URL is termed as the child URL. This has been done based on the assumption that it has a higher probability that if a page belongs to a particular domain, then the links embedded in that page or in the running text/paragraph of that page may belong to the same domain as of the domain of its parent page.

The above assumption leads to the calculation of the relevance score by including two values: a domain score that is calculated based on the terms that surround the URL in the containing paragraph and a Link Score that accounts for the relevance of the discovered URL to a domain based on the domain of its parent pages. Thus, the relevance score of the URL in domain D denoted as R-Score (URL, D) can be computed by:

$$R\text{-}Score(URL, D) = domain\_score(URL, D) + link\_score(URL, D) \quad (1)$$

The domain_score (URL, D) for a particular URL belonging to a domain D has been computed with the help of our work in [5], and the link_score (URL, D) for a URL belonging to domain D is computed with the help of domain information of its parent URL. The following subsections discuss the calculation of the domain and Link Scores in detail.

2.4.1 Domain Score

The Domain Score is used to quantify the effect of the terms contained in the running text or the paragraph of that new URL. So, its value is computed with the help of the various terms that surround the new URL and the domain definitions created by the machine learning tool. The domain definitions help to uniquely characterize a domain.

In order to compute the domain_score for a particular URL, the contents of its parent page (corresponding to the parent URL) have been checked, and the paragraph in which the new URL exists has been analyzed. Thus, for computing the value of the domain_score, the following steps have been followed:

1. **Running text identification**: For appropriate computation of the domain_score, the parent page has been first segmented into blocks or paragraphs for finding the URLs embedded in it. Each such block will have features like text, images, applets, tables that help in providing the best possible description of the URL. The paragraphs can easily be identified based on <p> tags associated with them in the HTML code of the web document.

2. **Tokenization**: As a next step, for each such paragraph that comprises a child URL, the various tokens or terms occurring in that paragraph have been identified.
3. **Stop-word Removal**: Based on a list of stop words, the terms or tokens that are having no meaning are eliminated from the list generated in step 2 above. For example: is, are, of, to, for, are some of the stop words.
4. **Domain Score computation**: The domain_score is computed on the basis of probability computation. From the remaining set of terms extracted from the paragraph, the number of terms occurring from each of the domains is identified. This is done by matching each of the extracted term against the key terms included in the domain definitions [5, 8] of each domain. The counts are then further used to compute the domain_score of the new/child URL which is given by the probability of the new/child URL belonging to a domain D. Thus,

$$\text{Domain\_score}(\text{URL}, \text{D}) = \frac{\text{Number of terms from domain D}}{\text{Total terms extracted pages}} \quad (2)$$

This domain_score is further used to compute the rank of the new/child URL on the basis of its R-Score.

Consider the web page (source: www.placetoseeindelhi.wordpress.com) shown in Fig. 3. The page provides a brief detail on the various places that one can visit in Delhi and belongs to the Travel domain.

Fig. 3 Example web page containing hyperlinks (*Source* www.placetoseeindelhi.wordpress.com)

<p>The Akshardham Temple is another impressive monument in Delhi. This creation is dedicated
to 11,000 artisans. It is located on the banks of glistering and pure Yamuna River. This stunning
architectural temple will mesmerize the travelers. The temple is too large; it takes one complete
day to see the entire places. One of the major highlights in Delhi is its provincial cuisines and
exotic dishes. Travelers can enjoy delicious foods in renowned restaurants and street stalls. Foods
available here is an influence of various culture and tradition. The taste and taste of Delhi cuisines
are simply irresistible.</p>

<p>Delhi is loaded with number of tourist attractions, sightseeing location, luxurious hotels and
other amenities to delight the travelers. The tourist attraction in Delhi includes Qutub Minar, Old
Fort, Jama Masjid, India Gate, Salimgarh fort and etc. The <strong><a
href="http://www.theweekendleader.com/Travel/1913/must-in-delhi.html">places to visit near
Delhi</a></strong> comprise Agra, Chandigarh, Mathura, Bharatpur, Corbett, Mandawa, etc.
Delhi is magical land and must to visit destination during the North India Travel. Its vibrant
culture, bustling stress, shopping centers and ancient monuments offer perfect combination of
modernity and ethnicity. Trip to Delhi is excellent way to spend your vacation in prime
destination by enjoying all modern amenities.</p>

Fig. 4 HTML tag structure of the web page in Fig. 3

The source code view of this parent web page is shown in Fig. 4 and refers to a
hyperlink places to visit near Delhi (illustrated by bold white text in the paragraph
in black). The Web Page Analyzer extracts the child URL "http://www.
theweekendleader.com/Travel/1913/must-in-delhi.html" embedded as this hyperlink in the parent page.

Now, in order to calculate the domain score of this new or child URL "http://www.
theweekendleader.com/Travel/1913/must-in-delhi.html", the running paragraph or
textual content is separated for tokenization and identification of the terms in the
comprising paragraph. Stop words are then removed from this obtained list of tokens
to extract the terms that seem useful and relevant for finding the domain_score.

Thus, the following terms have been extracted: *Delhi, loaded, number, tourist,
attractions, sightseeing, location, luxurious, hotels, amenities, delight, travelers,
Qutub, Minar, Jama, Masjid, India, Gate, Salimgarh, fort, places, visit, near, Agra,
Chandigarh, Mathura, Bharatpur, Corbett, Mandawa, magical, land, destination,
North, vibrant, culture, bustling, stress, shopping, centers, ancient, monuments,
perfect, combination, modernity, ethnicity, Trip, excellent, way, spend, vacation,
prime, enjoying, modern.*

The domain definitions for the purpose can be generated through the learning
process applied on initial data set. Neural networks [10] have been used for the
same in the proposed approach. The detailed process for learning the domain

definitions has been explained in our work in [5]. Considering that the following domain definitions have been generated for the Travel domain [5, 8]:

{Tour, Travel, visit, country, hotel, street, road, local, attraction, city, capital, metropolitan, world, nation, route, migration, price, vacation, railway, roadway, coming, leaving, source, arrival, economy, culture, heritage, airline, airway, reservation, age, passengers, adults, children, natural, return, largest, Asia, Africa, America}.

It has been found that the following seven terms, tourist, attraction, hotel, travel, visit, trip, culture (out of total 53 terms), occur in the domain definition of the Travel domain giving a value of

$$\text{domain\_score}(\text{URL},\text{Travel}) = \frac{7}{53} = 0.13 \tag{3}$$

The terms such as Agra, ancient, number, loaded that do not occur in a domain definition add a zero value to the count and thus do not modify the value of Domain Score for the URL in that domain. Thus, the value of Domain Score of the new URL in all other domains equals zero, i.e.,

$$\text{domain\_score}(\text{URL}, \text{Food}) = \text{domain\_score}(\text{URL}, \text{Books}) = 0 \tag{4}$$

Similarly, the value of Domain Score of a URL can be calculated for other domains. Also for cases where the same URL appears in more than one paragraph in the parent page, then the Domain Score of the URL is calculated for each of the paragraph block separately, and respective Domain Scores obtained for each paragraph are summed up to find the overall relevance of the URL in each of the various domains.

2.4.2 Link Score

The Link Score is used to predict the relevance of the new URL in the domain based on the domain information of its parent pages. The URL database contains the new/child URL that is extracted from the downloaded web page, the URL of its parent page, and the domain of that parent URL. Thus, it has been interpreted that the probability of the new URL as belonging to a domain of its immediate back-link is higher than the probability of it belonging to other domains.

Consider, for example, the new URL that is included in the URL database, and when the URL is searched for its parent pages, six different parent pages P_1, P_2, P_3, P_4, P_5, and P_6 have been identified. So, the new URL has a total of 6 back-links P_1, P_2, P_3, P_4, P_5, and P_6, and the domain information that is stored in the URL database for each of these six web pages is as Travel, Entertainment, Sports, Travel, Travel, and Entertainment, respectively. This data has been summarized as in Table 1.

The link_score of the new URL in each of the domains like Travel, Sports, and Entertainment is then calculated by finding the ratio of the number of parent pages

Table 1 Extracted back-links for the child URL and their associated domains

New/child URL	Immediate back-link →	P_1	P_2	P_3	P_4	P_5	P_6
	Domain of the immediate back-link →	Travel	Entertainment	Sports	Travel	Travel	Entertainment

in that domain and the total number of parent pages. Thus, the link_score of the new URL in TRAVEL domain is calculated as:

$$\text{Link\_score}(\text{URL}, \text{Travel}) = \frac{\text{Number of parent pages in travel domain}}{\text{Total number of parent pages}} = \frac{3}{6} = 0.5$$

(5)

Similarly,

$$\text{Link\_score}(\text{URL}, \text{Entertainment}) = \frac{2}{6} = 0.33 \tag{6}$$

$$\text{Link\_score}(\text{URL}, \text{Sports}) = \frac{1}{6} = 0.16 \ \& \ \text{Link\_score}(\text{URL}, \text{Food})$$
$$= 0 \ \& \ \text{Link\_score}(\text{URL}, \text{Food}) = 0 \tag{7}$$

The domain_score and link_score for the new URL are used to calculate the R-Score of the URL for different domains as per the Eq. (1). The calculation of the R-Score for the new/child URL is shown as an example in Table 2.

It may be observed that the Link Score considers the relevance of all the parent pages to the URL as the hyperlink is a reference to a child web page from all the parent pages that contain it.

Thus, the discovered URL will be added to the URL pool for Travel, Sports, and Entertainment domains with the respective R-Score of 0.63, 0.16, and 0.40.

Calculating the rank of each URL within a domain will help in focusing the crawl to the most popular and relevant links in a domain. URL Ranker after

Table 2 Relevance score calculation for new/child URL

Domain, D	domain_score (URL, D)	link_score (URL, D)	R-Score (URL, D)
Entertainment	0.07	0.33	0.40
Food	0	0	0
Travel	0.13	0.5	0.63
Sports	0	0.16	0.16

calculating the relevance scores adds the discovered URLs to its respective domain-specific URL pool. Thus, the URL Ranker finds the relevance of a new URL in each domain based on its Domain Scores and Link Scores.

3   Conclusion

It is very important for the crawler of the search engines to get the high-quality relevant webpages as rapidly as possible because the crawler cannot visit the entire web in a fairly reasonable amount of time. The approach seems of help to any focused crawler or domain-specific search services like vertical search engines which can utilize the domain information from the webpage to create numerous domain-specific data repositories. These data repositories being domain specific will act as a huge source of information and knowledge to the users. The proposed approach not only helps in designing efficient crawlers but also helps the user by providing better results with reduced response times.

References

1. A.P. Asirvhatam, K.K. Ravi, *Web Page Categorization Based on Document Structure* (International Institute of Information Technology, Hyderabad, India)
2. S. Lawrence, C.L. Giles, Searching the world wide web. Science **280**, 98–100 (1998) www. sciencemag.org
3. M. Diligenti, F. Coetzee, S. Lawrence, L. Giles, M. Gori, Focused crawling using context graphs, in *Proceedings of 26th International Conference on Very Large Data Bases, Cairo, Egypt*, pp. 527–534 (2000)
4. S. Chakrabarti, M. van den Berg, B. Dom, Focused crawling: a new approach to topic-specific web resource discovery. Comput. Netw. **31**(11–16), 1623–1640 (1999)
5. S. Gupta, K.K. Bhatia, Domain identification and classification of web pages using artificial neural networks, in *3rd International Conference on Advances in Computing, Communication and Control ICAC3, 18–19 Jan, 2013* organized by Fr. Conceicao Rodrigues College of Engineering, Mumbai, Proceedings by Springer, Berlin, pp. 215–226
6. S. Shibu, A. Vishwakarma, N. Bhargava, A combination approach for web page classification using page rank and feature selection technique. Int. J. Comput. Theor. Eng. **2**(6) (2010)
7. X. Qi, B.D. Davison, Knowing a web page by the company it keeps, in International conference on Information and knowledge management (CIKM), pp. 228–237 (2006)
8. S. Gupta, K.K. Bhatia, Optimal processing of search forms for hidden web extraction through a novel random ranking mechanism. Int. J. Inf. Retrieval Res. (IJIRR)
9. S. Gupta, K.K. Bhatia, Crawl part: creating crawl partitions in parallel crawlers, in *IEEE International Symposium on Computing and Business Intelligence, ISCBI August 2013*, held at Delhi organized by Cambridge Institute of Technology. IEEE Xplore
10. TheMathsWork, http://www.mathworks.com/products/neuralnet/

Performance Enhancement in Requirement Prioritization by Using Least-Squares-Based Random Genetic Algorithm

Heena Ahuja, Sujata and Usha Batra

Abstract Nowadays, software has become a part of greater significance in human life. The chief goal in today's environment is to prepare the software that can meet all the needs of stakeholders' increase in complexity of software boost in the requirement too. Many requirements are fulfilled in the given time duration or can be considered first to reduce the risk by the proper utilization of the software. In this way, gathering and prioritizing requirements can help in the successive development of the software. Requirement prioritization is one of the essential decision-making processes. Its main objective is to extract the suitable facts from the client, describing the requirements by the attributes. The ultimate purpose is to minimize the user decision-making effort and to increase the accuracy of the final requirements' ranking. A number of requirement prioritization techniques have been proposed till date but are not efficient performance wise. This piece of work presents a new technique which performs an enhancement in performance by using least-squares-based random genetic algorithm. The aim of this research is to assist engineers in requirement prioritization by reducing time and hence minimizing decision-making efforts. In the proposed work as the number of pairs increases, the distance is slightly increases. Select the initial population by analyzing some of the input partial orders, or it can be done randomly. The time also reduces. In future, to integrate genetic algorithm (GA) and value-oriented prioritization (VOP), take the benefit of both for the better outputs.

Keywords Requirement prioritization · Genetic algorithm · Analytical hierarchical process · Least-squares-based random genetic algorithm

H. Ahuja (✉)
Department of CSE & IT, NorthCap University, Gurgaon, India
e-mail: heenaahuja34@gmail.com

Sujata
Department of CSE & IT, Banasthali University, Gurgaon, India
e-mail: Sujata.bhutani@gmail.com

U. Batra
Department of CS & IT, GD Goenka University, Gurgaon, India
e-mail: Usha.batra@gdgoenka.ac.in

© Springer Nature Singapore Pte Ltd. 2018   251
B. Panda et al. (eds.), *Innovations in Computational Intelligence*, Studies in
Computational Intelligence 713, https://doi.org/10.1007/978-981-10-4555-4_17

1 Introduction

Requirement prioritization plays a very crucial role in software development process. We can define requirement prioritization as a process which prioritizes the requirement between the two choices that can help to select which one should be enforced first [1]. The prioritization of customer need is the key concern and is sort out by different prioritization techniques. First of all, some user's requirement is chosen on the basis of their priorities. The fundamental prioritization performs a vital act in decision making and also lessens the complexities because of in differentiable aims and needs [2]. Also, it can be named as complex decision-making process [3].

The aims can be achieved by determining the exact requirement. Mostly often, it is needed to prioritize the requirement so that the priority that has maximum requirement can be completed first.

Several methods are useful in requirement prioritization, for instance Analytical Hierarchical Process (AHP), Value-Oriented Prioritization (VOP), Numerical Assignment Technique (NAT), Binary Search (BS), Cumulative voting (CV), Multimedia Engineering Chance Understanding and Association (SERUM), Ranking, Arranging Game (PG), EVOLVE [4]. Such methods focus a number of factors of needs in different ways, for example, price, worth, chance, benefit [20]. That's why, there are plenty of objectives, which are discussed below:

- To analyze the essential and crucial need.
- To eliminate the irrelevant needs that will take place on accurate given duration.

According to these methods earlier Paolo Tonella and Angelo Susi both proposed an interactive genetic algorithm in which to integrate the Analytical Hierarchical Process and Genetic Algorithm [5]. In this approach, main problem is redundancy problem. To overcome this drawback, a new technique is proposed, i.e., least-squares-based random genetic algorithm. In the proposed work, as the number of pairs rises, the distance will also slightly rise. Select the initial population by analyzing one or more than one of the input partial orders, or it can be done randomly. The time also reduces [6]. Meanwhile, Value-Oriented Prioritization (VOP) and genetic algorithm (GA) take the advantages for best results. Additionally, we can integrate any of the techniques with partial swarm optimization.

2 Proposed Work

The purpose of this study is to enhance the performance for requirement prioritization by using least-squares-based random genetic algorithm. Select the initial population by analyzing many input partial orders or it can be done randomly. Minimize the error probability.

Proposed mathematical model is used for analyzing the performance distance with respect to the group of pairs. As the time decreases, distance will change again and again, and on increasing the number of pairs, distance coverage slightly increases and from this selection of the population randomly. The time also decreases.

3 LSRGA Empirical Study

Requirements $r1, r2, r3, r4\ldots rn$ would also consider as in population, and suppose we are taking two couple of requirements such as $R1$ and $R2$. One requirement has precedence over the other one can be stated by the extracted position, but it would be depend on high to low priority (e.g., $r1 \rightarrow r2$; or $r2 \rightarrow r1$ or $r3 \rightarrow r1$).

In the beginning, a precedence edge is presented in the obtained precedence graph.

Secondly, we would go for the number of generations for ($r2 \rightarrow r1$ or $r3 \rightarrow r1$). In next example, the client will be demanding to compare $(r1, r2)$, $(r1, r3)$ and $(r4, r5)$ and $(r1, r3)$, then after that the population of individuals through a set of fully ordered of requirements. According to our proposed technique, population and number of crossover mutates remove the duplicate computed requirement by using randomized method, or it can be formed by taking into account one or more of the input fractional orders so it would be cover as with (Figs. 1 and 2; Tables 1 and 2).

Where Pr as with crossover mutate and $r1, r2, r3$ with population disagree as with minimum pair mutation points are chosen for pairs, binary string from beginning of chromosome to the first crossover take place, i.e., $r2$ is copied from one $r1$ which is as per parent requirement, from the second parent requirement, i.e., of $r1$ has copied the part from the first to the second path crossover point.

Fig. 1 Priorities and dependencies

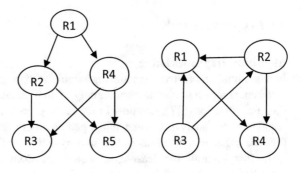

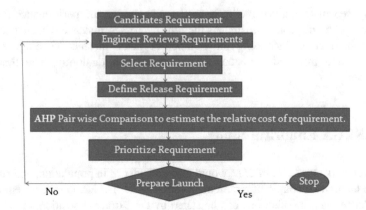

Fig. 2 Requirement prioritization using genetic algorithm

Table 1 Requirement with priority and dependencies

Requirement	Priority	Dependencies
r1	High	r2, r3
r2	Medium	r3
r3	Low	
r4	Medium	r1, r2
r5	Low	

Table 2 Prioritization requirement and related dependencies

Requirement	Priority	Disagreement
r1	Pr1	6
r2	Pr2	6
r3	Pr3	5
r4	Pr4	6
r5	Pr5	5

4   Proposed Model

Algorithm Implementation

In the initial interactive genetic algorithm, with a group of totally ordered of requirements to initialize the community of individuals through existing priorities (i.e., prioritizations) [7]. But now in the latest proposed approach, select the initial population by analyzing some input partial orders or it can be done randomly. Produce better initializations in this step by using the Greedy heuristics technique.

In this section, the least-squares-based random genetic algorithm (GA) is described that executes the new approach. Previously in interactive genetic algorithm, to deliver a recognized definition of the disagreement which acts an essential role when considering the fitness of an individual and selecting which pair-wise comparisons to obtain from the client.

Steps of Algorithm (a)
Input R: Group of needs
Input order 1 ... orders m: partial orders that refer priorities and constraints upon R.

1. Initialize population (pri)
2. Elicited pairs = 0
3. max elicited pairs = Max(default = 100)
4. threshold disagreement = TH(default = 5)
5. top population perc = 5%(PC)
6. eliord = null
7. for each pri in population do
8. calculate summation of disagreement in random way w.r.t. (ord1, ...ordk)
9. end for
10. while disagreement > threshold disagreement ^ execute time < total processing do
11. sample of population with bias toward randomized least square disagreement, e.g., sitting plan of examination
12. sort partial factor of population by increasing disagreement
13. mutate the best disagreement task
14. if min-disagreement did not decrease partially during last generation ^ randomized top population ^ elicited pairs < max elicited pairs using least square
15. eliminate duplicate pairs for floating point distance
16. eliord U elicit pair-wise comparison after elimination
17. mutate population in 3 cases:

    (i)   collect common disagreement for cut head
    (ii) swap the disagreement
    (iii) fill-in-tail operators

18. update all pairs
19. check n = 2,550,100 pairs for distance w.r.t. no. of iterations
20. update the maximum distance value of executed pairs in total no. of disagreements
21. end for
22. end while
23. Return Prmin, in least square concept.

This algorithm contains the pseudo-code of the least-squares-based random genetic algorithm. According to this algorithm, a group of requirements *R* is prioritized. A group of one or more than one partial orders (order 1, ... order *m*) are further input of the algorithm, resulted from the given requirement details (e.g., priority and dependencies).

5 Analysis

Algorithm (a) contains the pseudo-code of the least-squares-based random genetic algorithm. According to this algorithm, a group of requirements R is prioritized. A group of one or more than one partial orders (order 1, … order m) are further input of the algorithm, resulted as of the requirement details such as priority and dependencies. With a group of completely ordered of requirements, the population of individuals is considered and is initialized by the algorithm. Select the initial population by analyzing one or more than one of the input partial sequence or it can be done randomly.

Steps 1–3

On the priority bases, the population is initialized. Place only some important parameters that are used in the algorithm. Disagreement names for target level, by practically requesting to the client the highest number of pair-wise comparisons, and with the highest fitness the fraction of individuals that are taken for possible ties, determined by user interaction. The total optimization time constrains by the significant range that are used in the algorithm which is the greatest execution time (*time-out*). Primarily, with source to the input partial sequence with the help of their disagreement compute the fitness of the individuals.

Steps 4–5

If we are collecting 5 requirements, then we have to maintain crossover mutation after 5 requirements; after then, we have to again process for each population size. For each population after 5 requirements, generate cost value again for next thresholds for 2, 3, 4 till then 5. After proceeding 5 Population that having minimum cost value, i.e., prioritized value for GA.

Step 6

After execution, threshold value should be null; if value is not null, it means loop proceed for infinite requirement.

Steps 7–9

The main loop is entered in the algorithm. The new populations of individuals are produced as long as the disagreement is higher than the threshold value. Then, the algorithm stops, after allowing the execution time is extreme, and with the lowest disagreement with respect to the initial random sequence and the random sequence is obtained from the client, reports the individual. In the main loop, evolutionary iteration will take place.

Step 10

Start the while loop in which disagreement is higher than the threshold value, and the total processing time is greater than the execution time.

Steps 11–15

Selection (step 11) is the process which is to be performed first. According to this algorithm, any selection method can be used. In ascending order, best 5 cost values are placed. The resulting population is categorized by falling ties and disagreement is resulted for the proper individuals in the given population, when locally minimum disagreement is reached with available constraints. The user can be arranged in order to determine them whether there are many ties. Particularly, for pair-wise comparison, equally achieved individual disagreement pairs are submitted to the client; in this case, multiple times each pair is not shown to the user throughout the process and we should take care about that.

Steps 16–18

In the obtained precedence graph (*eliOrd*), the result of the comparison is stored. And eliminate the duplicate pairs. The population is growing between crossover and mutation for pair-wise comparison. For mutation, we use the collected common disagreement for cut head and swap the disagreement which consists of swapping the location in the mutated individual and selecting two R of requirements such as $R1$, $R2$. Swapping can be performed randomly and may either include adjacent or non-adjacent requirements and fill-in-tail operators. In this work, we examined only random selection.

In the chromosomes, a cut point of the first individual is selected by using fill-in-tail operator for crossover; with the missing requirements, store one of the heads or the tails, and fill in the tail according to the order obtained in the next individual to be crossed throughout ordered them.

Chromosomes from phase to phase that are already a part of the new population created may produce the crossover, and mutation operators are described earlier. It becomes a problem in degenerate cases. To overcome such a problem, the generation of chromosomes is limited which is already available in the population being created and measure of population diversity is introduced. Until the population range passes a predefined threshold value, then the crossover and mutation are applied repeatedly. Update all pairs.

Steps 19–22

After that selection of the finest individuals to decide the fitness, survey is to be utilized. After study the disagreement that takes into the form of an account of the initial partial orders with the selected precedence's achieved through following user connections. The distance will increase as the total number of pairs increases. Finally, return minimum error probability (which based on least square) for our technique.

6 Results

According to the results, least-squares-based random genetic algorithm with respect to time and distance is proposed. The x-axis shows the range of time. Distance has been shown in the y-axis. Now, check order according to each pair p1, p2, p3 such that 25, 50, 100, respectively. Whenever decreasing the time, distance will change again and again. Whenever we increase the number of pairs, distance coverage slightly increases. The ordering pairs were randomly considered, since the least-squares-based random genetic algorithm involves nondeterministic steps. Also, it removes the redundancy problem. And minimize the error probability. According to Fig. 3, performance comparison distance with respect to grouping in terms of requirement for 25 pairs increases; after that according to Fig. 4, it will get slightly decrease; and then from Fig. 4, performance comparison distances with respect to grouping in terms of requirement for 100 pairs are growing up. And it gives the better result as compared to the previous technique. And also, it gives the accurate result (Fig. 5; Tables 3, 4, and 5).

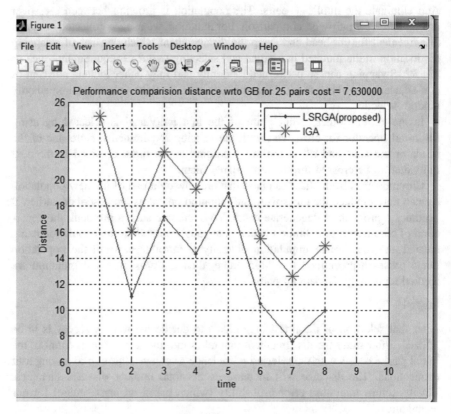

Fig. 3 Performance comparison distance with respect to grouping in terms of requirement for 25 pairs

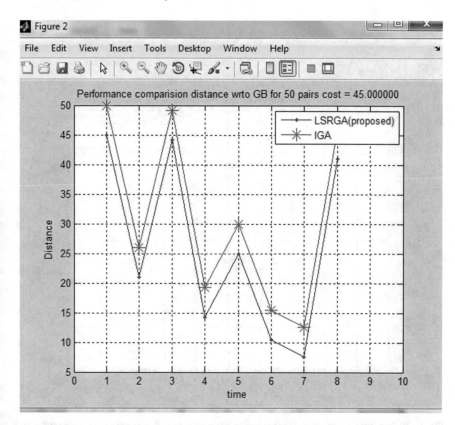

Fig. 4 Performance comparison distance with respect to grouping in terms of requirement for 50 pairs

In Fig. 6, the distance is getting increases as the number of pairs increases. So in the last graph, the 100 pairs will give the better result as compared to the earliest algorithm. As the distance increases, order will get better result. As the number of pairs increases, the distance will get increases (Table 6).

7 Conclusion

Here, in order to prioritize the requirement, we have mentioned nine techniques; out of all these, VOP (value-oriented prioritization) is regarded as the best since it gives non-erroneous results and consists of easy-to-use method. Moreover, it can help us to judiciously handle more requirements such as EVOLVE and SERUM; both

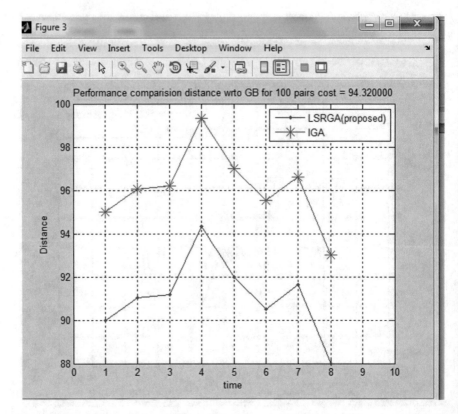

Fig. 5 Performance comparison distance with respect to grouping in terms of requirement for 100 pairs

Table 3 LSRGA and interactive genetic approach for 25 pairs

S. No.	Distance (IGA)	Time (s)	Distance (LSRGA)
1	25	1	20
2	16	2	11
3	22	3	17
4	19.1	4	1
5	24	5	19
6	15.6	6	1
7	13.1	7	7.8
8	15	8	10

Table 4 LSRGA and interactive genetic approach for 50 pairs

S. No.	Distance (LSRGA)	Time (s)	Distance (IGA)
1	45	1	50
2	21	2	26
3	44	3	44
4	15	4	20
5	25	5	25
6	10	6	15
7	8	7	13
8	41	8	47

Table 5 LSRGA and interactive genetic approach for 100 pairs

S. No.	Distance (LSRGA)	Time (s)	Distance (IGA)
1	90	1	95
2	91	2	96
3	90.2	3	96.3
4	94.5	4	98.7
5	92	5	97
6	91	5	95.7
7	91.7	7	96.6
8	00	8	93

calculate the advantages while implementing the needs, and at the same time, EVOLVE is unable to estimate the cost factor. One can easily take into account the 3 risk factors: SERUM, EVOLVE and VOP. VOP reduces the consumption of time, and EVOLVE examines the precedence and coupling constraints among all the requirements. In the coming time, we can merge VOP and genetic algorithm to get better yield. This piece of work presents a new technique which performs an enhancement in performance by using least-squares-based random genetic algorithm as compared to the earlier interactive genetic approach. In LSRGA, as the number of pairs increases, the distance also increases. It will give the better results. If we increase the number of pairs, then it will provide much better results and accuracy.

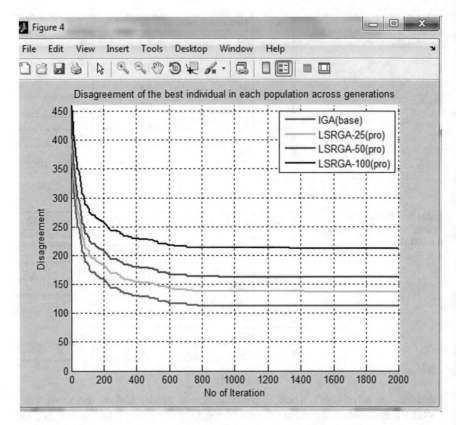

Fig. 6 Each population's disagreement of best individual across number of generations

Table 6 Each population's disagreement of the best individual

S. No.	No. of iteration	Disagreement of IGA	Disagreement of LSRGA-25	Disagreement of LSRGA-50	Disagreement of LSRGA-100
1	200	140	170	200	250
2	400	130	150	180	230
3	600	130	147	170	220
4	800	130	140	160	220
5	1000	120	140	160	210
6	1400	110	140	160	210
7	1600	110	140	160	210
8	1800	110	140	160	210

References

1. A. Iqbal, F.M. Khan, S.A. Khan, A critical analysis of techniques for requirement prioritization. Int. J. Rev. Comput. **1**, 8–18 (2009)
2. M. Khari, N. Kumar, Comparisons of techniques of requirement prioritization. J. Glob. Res. Comput. Sci. **4**(1), 38–43 (2013)
3. D. Greer, D. Bustard, T. Sunazuka, Prioritization of system changes using cost-benefit and risk assessments, in *Proceedings of the Forth IEEE International Symposium on Requirements Engineering*, (SERUM) (1999b), pp. 180–187
4. J. Azar, R.K. Smith, D. Cordes, Value oriented requirements prioritization in a small development organization. IEEE Softw. 32–73 (2007)
5. P. Tonella, A. Susi, *Using Interactive GA for Requirements Prioritization* (Information and Software Technology, 2012), pp. 1–15
6. S. Heena Ahuja, G.N. Purohit Understanding Requirement Prioritization Techniques, in *IEEE International Conference Computing Communication and Automation (ICCCA2016)*, Vol. 2 (2016)
7. D. Greer, G. Ruhe, Software release planning: an evolutionary and iterative approach. J. Inf. Softw. Technol. **46**(4), 243–253 (2004)
8. H.F. Hofmann, F. Lehner, Requirement engineering as a success factor in software projects. IEEE Softw. 58–66 (2001)
9. J. Kalorsson, K. Ryan, prioritizing requirements using a cost-value approach. IEEE Softw. **14** (5), 67–74 (1997)
10. J. Karlsson, Software requirements prioritizing, in *Proceedings of 2nd International Conference on Requirements Engineering* (ICRE'96), April 1996 (1996), pp. 110–116
11. J. Karlsson, C. Wohlin, B. Regnell, An evaluation of methods for prioritizing software requirements. Inf. Softw. Technol. 939–947 (1998)
12. D. Leffingwell, D. Widrig, *Managing Software Requirements: A Use Case Approach*, 2nd edn. (Addison-Wesley, Boston, 2003)
13. P. Berander, A. Andrews, Requirements prioritization. Inf. Softw. Technol.: Part 1 69–94 (2005)
14. T.L. Saaty, *The Analytic Hierarchy Process: Planning Priority Setting Resource Allocations* (McGraw-Hill Inc., 1980)
15. F. Fellir, K. Nafil, R. Touahni, *System Requirements Prioritization Using AHP* (IEEE, 2014), pp. 163–167
16. M. Ramzan, I. Aasem, M. Arfan Jaffar, Overview of existing requirement prioritization techniques, in *International Conference on Information and Emerging Technologies (ICIET)*, June 2010, Karachi, Pakistan (2010)
17. R. Beg, Q. Abbas, R.P. Verma, An *Approach* for *Requirement Prioritization* (IEEE Computer Society, 2008)
18. N. Kukreja, B. Boehm, S. Padmanabhuni, S.S. Payyavula, *Selecting an Appropriate Framework for Value-Based Requirements Prioritization* (IEEE Software, 2012), pp. 303–308
19. A. Hambali, S. Mohd, N. Ismail, Y. Nukman, Use of analytical hierarchy process (AHP) for selecting the best design concept. J. Technol. **49**(1), 1–18 (2008)
20. S. Valsala, A.R. Nair, Requirement prioritization in software release planning using enriched genetic algorithm. Int. J. Appl. Eng. Res. (IJAER) **10**, 34652–34657 (2015)

Regression Testing Using User Requirements Based on Genetic Algorithm

Kuldeep Kumar and Sujata

Abstract Regression is done in the two circumstances one if programming has been changed (in light of bug fixes or adding additional usefulness or erasing existing usefulness), and second, if the environment changes, still we will do regression testing. Regression testing will be directed after any bug settled or any usefulness changed. Test suite/case minimization procedure addresses this issue by expelling repetitive experiments and test suite prioritization procedure by testing test cases in a request that upgrade the effectiveness of achieving a few execution criteria. This paper displays another methodology for regression testing joining these two strategies. The methodology is to first minimize the test suites by utilizing covetous methodology and after that organize these minimized test cases and prioritize them by utilizing genetic algorithm (GA). Proposed approach underpins analyzer by minimizing and prioritizing the test suite while guaranteeing all the necessary scope and least execution cost and time using user requirements.

Keywords Regression testing · Test case prioritization · User requirements
Genetic algorithm

1 Introduction

Regression testing is a sort of programming testing that checks that product already created tried still performs effectively after it was changed or interfaced with other programming. Changes may incorporate programming upgrades, patches, setup changes, and so on. These ranges may incorporate utilitarian and non-practical

K. Kumar (✉)
Department of Computer Science, The NorthCap University, Gurgaon, India
e-mail: kuldeep14csp024@ncuindia.edu

Sujata
Department of Computer Science and Engineering, GD Goenka University,
Gurgaon, India
e-mail: sujata@gdgoenka.ac.in

© Springer Nature Singapore Pte Ltd. 2018   265
B. Panda et al. (eds.), *Innovations in Computational Intelligence*, Studies in
Computational Intelligence 713, https://doi.org/10.1007/978-981-10-4555-4_18

zones of the framework. Retesting (additionally called confirmation testing) when testing is done to affirm that the bug which we reported before has been settled in new form is called retesting; however, regression testing implies testing to guarantee that the fixes have not acquainted new bugs with be show up. Regression testing is a costly and extravagant procedure in which various methodologies enhance viability as it might account 70% of the aggregate expense (in an study).

For Example there are three sorts and modules in the Project named Admin Module, Personal Information, and Employment Module and accept bug happens in the Admin Module like on Admin Module existing User is not prepared to login with genuine login accreditations so this is the bug. In no time Testing bunch sends the above—determined Bug to the Development gathering to adjust it and when progression bunch settles the Bug and hand over to Testing bunch than testing bunch watches that changed bug does not impact the remaining handiness of exchange modules (Admin, PI, Employment) moreover the helpfulness of the same module (Admin) so this is known as the technique of backslide testing done by Software Testers.

1.1 Re-arrange All

It is one of the traditional and customary frameworks for regression programming testing in which each and every investigation in the current experiments/suites is retuned. This methodology is not reasonable most of time as it requires additional time, cost, and spending arrangement.

1.2 Regression Test Case Selection and Reduction

Here, RTS licenses us to block a rate of the investigations and tests. RTS is valuable just if the cost of selecting a bit of the examinations is not precisely the cost of executing the complete test suite. In this strategy, simply bit of investigations in test suite is rerun. RTS systems are further assembled into three orders: [1] Coverage Technique, Minimization Technique, and Safe Technique.

1.3 Test Case Prioritization

Test case prioritization frameworks sort out tests in a most supportive solicitation, thusly making the testing handle all the more convincing. There are 18 different trial prioritization strategies [2] numbered P1–P18 which is isolated into three social events: comparator techniques, statement-level techniques, and function-level

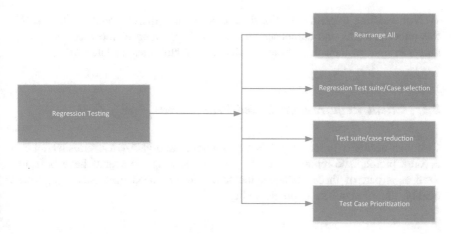

Fig. 1 Techniques used for regression testing

techniques. With a specific end goal to build cost adequacy and productivity, prioritization methodology is utilized to organize the experiments by revamping (Fig. 1).

2 Related Work

The time-obliged experiment/time-constrained TC prioritization issue can be powerful and diminished to the NP-complete zero/one knapsack issue/problem [3]. This can regularly be productively approximated with a GA heuristic search procedure. Genetic algorithm has been viably utilized as a part of other programming building and programming dialect issues and making time-obliged test prioritizations utilizing prerequisites variables and procedure that organizes relapse test suites so that the new requesting:

1. Will dependably keep running inside a given time limit.
2. Will have the most elevated conceivable potential for serious imperfection identification in view of inferred scope data and necessities.

In rundown, the imperative commitments of this overview are as per the following:

(i) a GA-based strategy to organize a relapse test suite that will be keep running inside a period compelled execution environment.
(ii) an exact assessment of the viability of the subsequent prioritizations in connection with (i) GA-delivered prioritizations utilizing distinctive client prerequisite parameters.

With existing experiment prioritization systems inquired about in 1998–2015, this paper presents and composes another "4C" grouping of those current procedures, in light of their prioritization calculation's attributes, as takes after.

2.1 Client Requirement-Based Procedures

Customer necessity-based strategies are techniques to organize test cases taking into account prerequisite reports. Likewise, numerous weight elements have been utilized as a part of these methods, including custom need, necessity many-sided quality, and prerequisite unpredictability.

2.2 Coverage-/Scope-Based Procedures

Coverage-based strategies are techniques to organize test cases in light of scope criteria, for example, prerequisite scope, all out necessity scope, extra prerequisite scope, and articulation scope.

2.3 Cost-Effective/Savvy-Based Strategies

Cost successful-based procedures are techniques to organize test cases taking into account costs, for example, expense of investigation and expense of prioritization.

2.4 Chronographic History-Based Strategies

Chronographic history-based procedures are techniques to organize test cases taking into account test execution history.

Before managing prioritization calculations, the issue connected with experiment prioritization requires understanding which is characterized as takes after.

3 Problem Given

Tc, a test suite, PTc, the arrangement of changes of Tc, and f, a capacity from PTc to the genuine numbers [2].

3.1 Problem

Find Tc' ∈ PTc
such that

$$(\Lambda Tc'')(Tc'' \in PTc)(Tc'' \neq Tc')$$

$$[f(Tc') \geq f(Tc'')]]$$

In this definition, PTc speak to the arrangement of all conceivable prioritizations (orderings) of Tc, and f is a function that, connected to any such requesting, yields a recompense esteem for that requesting [2]. (F or effortlessness, and without loss of all-inclusive statement, the definition accept that higher grant qualities are desirable over lower ones.)

4 TCP Approach Utilizing GA

The genetic calculation is a transformative calculation and populace-based pursuit technique. The determination happens from the accessible populace utilizing fitness capacity; genetic administrators are connected to acquire an ideal arrangement, took after termination. In genetic algorithm, populace of chromosome is defined by various codes, for example, parallel, real numbers and so on genetic operators (i.e., determination/selection, hybrid/crossover, change/mutation) is connected on the chromosome with a specific end goal to discover more fittest chromosome. The wellness and fitness of a chromosome are characterized by a reasonable target capacity. As a class of stochastic strategy, genetic algorithm is not quite the same as an irregular search.

The progressions of hereditary calculation are as follows:

1. Generate/produce population (chromosome).
2. Assess the fitness of produced population.
3. Apply determination/selection for individual.
4. Apply hybrid/crossover and transformation/mutation.
5. Assess and imitate the chromosome.

5 Implementation and Results

In this section, we present technique to make regression testing efficient using genetic algorithm in MATLAB tool. Let us say a program, P, has a test suite T in the form of test cases representing the requirements each test case is covering as shown in Table 1.

Table 1 Test cases representing user requirements

Test case	User requirement						
	Ur1	Ur2	Ur3	Ur4	Ur5	Ur6	Ur7
Tc1	1	0	1	1	0	0	0
Tc2	0	1	1	1	1	0	0
Tc3	1	1	1	1	0	0	0
Tc4	0	0	0	0	1	0	1
Tc5	1	0	0	1	0	1	0
Tc6	1	0	1	0	0	0	0
Tc7	1	0	1	1	0	1	0
Tc8	0	1	0	0	0	1	0
Tc9	0	0	0	0	0	0	0
Tc10	0	1	0	1	1	1	1

The trial experiment is actualized in MATLAB. In this approach, the execution time of each experiment is additionally examined. The flaw measuring system utilized as a part of deficiency scope is based on trying method. In this case, there are experiments framing test suite.

$$TS = \{Tc1, Tc2, Tc3, Tc4, Tc5, Tc6, Tc7, Tc8, Tc9, Tc10\}.$$

Similarly, the announcements secured by the experiments are signified as statement covered (SC) = {Ur1, Ur2, Ur3, Ur4, Ur5, Ur6, Ur7} (Ur: User requirements). We are characterizing another way to deal with minimize and afterward needs the experiments utilizing genetic algorithm calculation. For this reason, we have utilized the MATLAB test system.

As greedy, prioritization algorithm may not generally pick the optimal/ideal tests requesting. To see this, assume a system contains four blames, and assume our test suite for that program contains three tests that distinguish those shortcomings as appeared in Table 2.

Table 2 Test cases after minimization

Test cases after minimization	Ur1	Ur2	Ur3	Ur4	Ur5	Ur6	Ur7
Tc1	1	0	1	1	0	0	0
Tc2	0	1	1	1	1	0	0
Tc5	1	0	0	1	0	1	0
Tc6	1	0	1	0	0	0	0
Tc7	1	0	1	0	0	0	0
Tc8	0	1	0	1	0	1	0
Tc9	0	0	0	0	0	0	0
Tc10	0	1	0	1	1	1	1

Minimized test cases/suites
Test cases eliminated = 3
Test case reduced to = 7

Table 3 Minimum test case with parameter

Parameters	Values
No. of test suites/cases	7
Fitness function applied	Minimization and prioritization
Creation/generation size	100
Testing cost	0–1
Crossover	PMX

Table 4 Severe faults in test suite

Test cases	Severe fault cases			
	1	2	3	4
Tc3	X	X		
Tc4	X			X
Tc10		X	X	X

Table 5 Prioritization sequence of test suites

Test cases after prioritization	Ur1	Ur2	Ur3	Ur4	Ur5	Ur6	Ur7
Tc2	0	1	1	1	1	0	0
Tc7	1	0	1	0	0	0	0
Tc9	0	0	0	0	0	0	0
Tc1	1	0	1	1	0	0	0
Tc5	1	0	0	1	0	1	0
Tc8	0	1	0	1	0	1	0
Tc6	1	0	1	0	0	0	0

For prioritization the minimized test suites an arbitrary expense is doled out to every test. Organized outputs for this irregular task are appeared in Table 5. The premise parameters are characterized while performing GA for test case prioritization (Tables 3 and 4).

5.1 MATLAB Implementation

Test cases are taken: [1 0 1 1 0 0 0; 0 1 1 1 1 0 0; 1 1 1 1 0 0 0; 0 0 0 0 1 0 1; 1 0 0 1 0 1 0; 1 0 1 0 0 0 0; 1 0 1 0 0 0 0; 0 1 0 1 0 1 0; 0 0 0 0 0 0 0; 0 1 0 1 1 1 1].

Total no. of requirements are taken for execution: [1 1 1 1 1 1 1 1 1 1].

As shown below (Figs. 2 and 3).

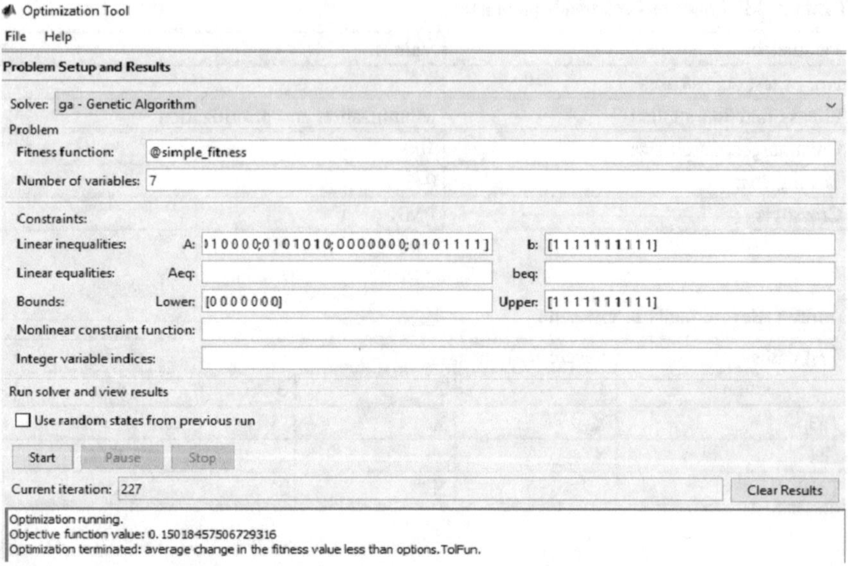

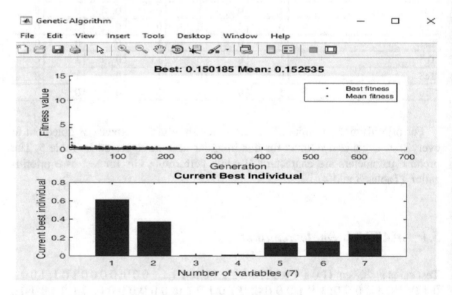

Fig. 2 Implementation using MATLAB optimization tool

Fig. 3 Graph 1: shows user-prioritized sequence of requirements

6 Evaluation of Approach

Consider the example of program with number of fault detected for particular test cases and severity value against those faults for the evaluation of this approach.

Illustration (Table 6) for doling out the need which can be arranged by client and improvement/development group. Weights to every component are appointed by the advancement/development group as indicated by the venture. Allocated all out weight (1.0) is separated among the PFs. For each prerequisite, Eq. (1) is utilized to ascertain a weighted prioritization (WP) component that measures the significance of testing a necessity prior [4].

$$\text{Wp} = \sum (\text{pf value} * \text{Pf weight}) \quad \text{where pf value is 1 to n} \qquad (1)$$

Wp defines the weighted prioritization for the four prerequisites figured utilizing Eq. (1) [4]. The aftereffects of the table demonstrate the prioritization of experiments for the four prerequisites as takes after: R1, R4, R3, and R2. The Wp will change with an adjustment in variable weights and component values.

Every deficiency is relegated seriousness/severity measure (SM) on a 10-point scale as demonstrated as follows:

- Complex (Severity 1): SM estimation of 9–10,
- Moderate (Severity 2): SM of 6,
- Low (Severity 3): SM of 4,
- Very Low (Severity 4): SM of 2.

To compute the total percentage of fault detected (TSFD), we use severity measure(SM) of every deficiency [5, 6]. Once the issue has been recognized, then we relegate some severity measure to every flaw as indicated by prerequisite weights, to which it is mapped. Absolute severity of faults detected (ASFD) is the summation of seriousness/severity measures of all deficiencies distinguished for an item.

$$\text{TSFD} = \Sigma \, \text{SM(severity measure)} \quad [\text{where } I = 1 \text{ to } n] \qquad (2)$$

Table 6 Example program

Req. factors	R1	R2	R3	R4	Weights
Business value based	7	4	10	8	0.35
Requirement volatility	10	5	4	9	0.30
Implementation complexity	10	8	5	5	0.25
Fault proneness of requirements	4	6	5	7	0.10
Wp	8.05	5.25	6.4	7.75	1.00

Table 7 Test prioritization

Tc2	Tc7	Tc9	Tc1	Tc5	Tc8	Tc6
2 faults	3 faults	5 faults	4 faults	2 faults	3 faults	2 faults
17 severity faults	14 sev.	11 sev.	9 sev.	7 sev.	5 sev.	3 sev.

This condition demonstrates TSFD for an item where n speaks to all out number of shortcomings recognized for the item [6]. Initial step is to manufacture the prioritization lattice in light of the variable estimation of the necessity. At that point, delineate experiments against each necessity. Execute the tests in light of the doled out need and break down the outcomes taking into account flaw seriousness/fault severity. The prioritization is shown in Table 7: Tc2 > Tc7 > Tc9 > Tc1 > Tc5 > Tc8 > Tc6.

On a live project, analysis is done using this algorithm. Existing approaches are compared to this, and it found more effective than others as shown in Figs. 4 and 5 as to TS (total severity)-R, TS (total severity using optimization)-O, TS (total severity)-F.

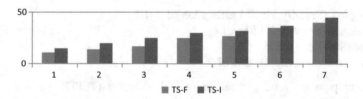

Fig. 4 Graph 2: *X*-axis: test suite fraction and *Y*-axis: percentage of TFS (total fault severity using requirement prioritization weight)

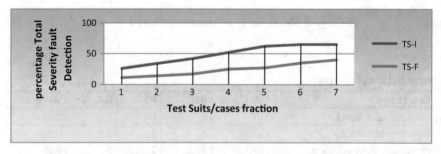

Fig. 5 Graph 3: This graph shows that when 7% of testing fraction executed, maximum number of fault severities founds which save the cost and time by achieving maximum client requirements

7 Conclusion

In this paper, a methodology and approach join the minimization strategy and prioritization procedures of regression testing. Genetic algorithm is best decision for prioritization as utilizing this algorithm genuinely substantial number of time we will get ideal and optimal arrangement which can tackle substantial number of test cases/experiments in productive time to find severe faults in less cost and less time. The algorithm is tackled and solved manually and is a stage toward test automation. In future, a mechanization instrument is to be created to actualize the proposed work.

References

1. P. Tonella, A. Susi, F. Palma, Using/utilizing interactive GA for requirements prioritization (GA), in *2nd International Conference Based Software Engineering*
2. G. Rotherrmell, R.H. Untech, C. Chui, M.J. Harrold's, Test/experiment case prioritization (TCP). Technical Report GIT-9-28, School of Figuring, Georgia Foundation of Innovation, Dec 1999
3. G.M. Kapfhammer, R.S. Roos, M.L. Soffa, K.R. Walcott, Time-aware test suite prioritization, in *Proceedings of the 2006 International Symposium on Software Testing and Analysis*, ISSTA'06, 17–20 July 2006, Portland, Maine, USA (2006)
4. H. Sri-kanth, L. Williams, J. Osborne, System test case prioritization of new and regression test cases, in *Proceedings of the Seventh Worldwide Workshop on Economics Driven Programming Building Research*, May 2005, IEEE, pp. 64–73
5. V. Kumar, Sujata, M. Kumar, Requirement based test case prioritization utilizing genetic algorithm. Int. J. Comput. Sci. Technol. **1**(1) (2010)
6. Sujata, M. Kumar, V. Kumar, Requirements based test case prioritization utilizing genetic algorithm. Int. J. Comput. Sci. Technol. **1**(2) (2010)
7. K.F. Man (IEEE Member), K.S. Tang, S. Kwong (IEEE Member), Genetic algorithms—concepts and applications. IEEE Trans. Ind. Electron. **43**(5) (1996)
8. V. Kumar, Sujata, M. Kumar, Test case prioritization utilizing/using fault severity. Int. J. Comput. Sci. Technol. **1**(1) (2010)
9. G. Roth, R. Untch, C. Chu, M. Harr, Test case prioritization/(TCP). IEEE-Trans. Softw. Eng. **27**, 929–948 (2001)
10. A. Malishevsky, S. Elbaum, G. Rothermel, Test case prioritization: a family of empirical studies. Trans. IEEE Softw. Eng. **28** (2002)
11. M.I. Kayes, *Test Case Prioritization for Regression Testing Based on Fault Dependency* (IEEE, 2011)
12. M. Harman, Z. Li, R.M. Hierons, Search algorithm for regression TC prioritization. IEEE Trans. Softw. Eng. **33**(4) (2007)
13. W. Junn, Z. Yann, J. Chen, Test case prioritization technique taking into account/based on genetic algorithm. IEEE (2011)
14. B. Suri, G. Duggal, *Understanding Regression Testing Techniques* (COIT, India, 2008)
15. M. Thillaikarasi, K. Seetharaman, Effectiveness of test suites/case prioritization techniques based on regression testing. Int. J. Softw. Eng. Appl. (IJSEA) **5**(6) (2014)
16. G.N. Purohit, Sujata, A schema support for selection of test case prioritization techniques. Int. J. Softw. Eng. Appl. (IJSEA) **5**(6) (2014)

17. Sujata, N. Dhamija, TCP using model based test case/suites dependencies: a survey. Int. J. I&C Technol. **4–10** (2014)
18. P. Berander, A. Andrews: Requirements prioritization, in *Engineering and Overseeing Programming Requirements*, ed. by C. Wohlin, A. Aurum (Springer, Berlin, 2005)
19. N. Sharma, Sujata, G.N. Purohit, Test case prioritization techniques. An empirical study. IEEE (2014)
20. L. Bhambhu, Jyoti, An efficient GA for fault base regression TCP. Int. J. Adv. Res. Comput. Commun. Eng. (2015)
21. https://www.google.co.in/

Rainfall Forecasting Using Backpropagation Neural Network

Sudipta K. Mishra and Naresh Sharma

Abstract Rainfall is an important hydro-climatic variable on which crop productivity, aridness etc. depend. Different time series analysis techniques, e.g. ARIMA, HWES etc. are typically used to predict rainfall. In this paper, authors applied different neural network techniques (ANN) for studying the rainfall time series in Burdwan district of West Bengal, India, by using past 15 years' data. Then, efficiency of different ANN schemes to predict the rainfall was compared and best scheme was selected. All the calculation works were done using ANN model in MATLAB R2013a.

1 Introduction

Predicting rainfall is one of the most challenging tasks in modern science as climate is a complex phenomenon having nonlinear characteristics. Artificial neural network can be used in such cases as it has the ability to model nonlinear complex relationship amongst input and output datasets [1–4]. This study explores the suitability of forecasting rainfall using two backpropagation neural network (BPNN) training algorithms namely LMBP (Levenberg-Marquardt backpropagation) and BRBP (Bayesian regulation backpropagation). It also compares the models and concludes the best model to predict rainfall.

S.K. Mishra (✉) · N. Sharma
GD Goenka University, Gurgaon, India
e-mail: sudipta.mishra@gdgoenka.ac.in

N. Sharma
e-mail: naresh.sharma2006@gmail.com

© Springer Nature Singapore Pte Ltd. 2018
B. Panda et al. (eds.), *Innovations in Computational Intelligence*, Studies in
Computational Intelligence 713, https://doi.org/10.1007/978-981-10-4555-4_19

2   Dataset Used

According to the IPCC 2014 report [5], climate change is going to have negative effects particularly for tropical cereal crops like rice [6]. Among Indian states, West Bengal holds the top position in rice cultivation with highest production from the high productive southern district in the fertile Gangetic plane namely Bardhaman, which is commonly known as the 'Rice bowl of Bengal'. Coming to rainfall, Fig. 1 reveals that over the 20 years, there has been a gradual decline in the average annual rainfall in Bardhaman.

Interestingly, the monthly trend of maximum and minimum temperature shows that in the last 15 years, summer (April–July) temperature has increased whereas winter (October–February) temperature has reduced compared to the previous years (Fig. 2). This implies that in Bardhaman the maximum temperature is becoming extreme alongside of a general increase in temperature.

In this study, we used temperature as an exogenous input to forecast rainfall using dataset from Bardhaman.

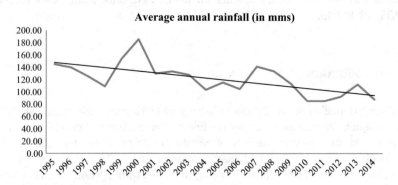

Fig. 1 Trend in average annual rainfall in Bardhaman. *Source* Prepared by the authors with data from Directorate of Agriculture, Government of West Bengal

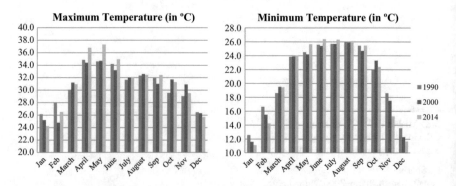

Fig. 2 Trend in maximum temperature versus minimum temperature in Bardhaman. *Source* Prepared by author with data from Directorate of Agriculture, Government of West Bengal

3 Research Methodology

The nonlinear autoregressive network with exogenous inputs (NARX) is a recurrent dynamic network, with feedback connections enclosing several layers of the network. The NARX model is based on the linear ARX model, which is commonly used in time series modelling [7].

Mathematically, NARX model is defined as

$$y(t) = f\big(y(t-1), y(t-2), \ldots, y(t-n_y), u(t-1), u(t-2), \ldots, u(t-n_u)\big)$$

where the next value of the dependent output $y(t)$ is regressed on previous values of the output and previous values of an independent (exogenous) input.

In this paper, comparison of two training algorithms such as Levenberg-Marquardt backpropagation and Bayesian regulation backpropagation are discussed.

(a) **Levenberg-Marquardt backpropagation**

Levenberg-Marquardt backpropagation is a network training algorithm that updates weight and bias values according to Levenberg-Marquardt optimization. This algorithm is one of the fastest backpropagation algorithms, and is highly recommended as a first-choice supervised algorithm, although it does require more memory than other algorithms [8].

(b) **Bayesian regulation backpropagation**

Bayesian regulation backpropagation is a network training function that updates the weight and bias values according to Levenberg-Marquardt optimization. It minimizes a combination of squared errors and weights, and then determines the correct combination so as to produce a network that generalizes well [9]. During training, the inputs to the network are real and training process will be more accurate. The neural network training can be made more efficient by normalizing the data into interval [0, 1]. The best neural network with less value of MSE is assumed.

4 Data Analysis

In this paper, backpropagation training algorithms, i.e. Levenberg-Marquardt and Bayesian regulation, were used to predict rainfall time series values, $y(t)$, from the past value of that time series. For exogenous variables, following three cases were considered:

1. Minimum temperature time series (as first case).
2. Maximum temperature time series (as second case).
3. Minimum and maximum temperature time series (as third case).

In the study, NARX with different number of hidden layers, different delay (d) and one output neuron were used [10]. The learning rate 0.001 is used in

Levenberg-Marquardt backpropagation algorithm to train the data. The experiment on the model was repeated by using the number of nodes from 2 to 10 and number of tapped delays from 2 to 4. The MSE value was observed for all networks of all three cases.

From Tables 1, 2, 3, 4, 5 and 6, it is clear that the lowest MSE value is 0.006 with five nodes in hidden layer and four tapped delay lines (first case), 0.0041 with eight nodes in hidden layer and two tapped delay lines (second case) and 0.002 with four nodes in hidden layer and four tapped delay lines (third case). Lowest MSE is supported by high value of R value 0.8277, 0.8451 and 0.8302, respectively.

MSE graph of neural network model produced during training by Levenberg-Marquardt backpropagation training algorithm is shown in Fig. 3. The training steps ended at the end of seventeenth epochs as the validation errors increased. At the end of neural network model training, MSE (mean square error) is obtained as 0.0072687 as shown in Fig. 4, whereas 'R' values are above 0.76. This implies that the output produced by the network is closer to the target and that the model is suitable (Fig. 5).

Table 1 Levenberg-Marquardt backpropagation model selection for first case (MSE)

No. of nodes	No. of delay		
	2	3	4
2	0.0075	0.0074	0.0092
3	0.021	0.0179	0.013
4	0.0136	0.0128	0.0072
5	0.0083	0.1011	**0.006**
6	0.0061	0.0115	0.0131
7	0.0211	0.012	0.0096
8	0.0091	0.0073	0.0116
9	0.0092	0.0084	0.0087
10	0.0095	0.0097	0.0085

Table 2 Levenberg-Marquardt backpropagation model selection for first case (R)

No. of nodes	No. of delay		
	2	3	4
2	0.8411	0.7449	0.7662
3	0.6802	0.7621	0.7979
4	0.6721	0.6939	0.7487
5	0.7464	0.7237	**0.8277**
6	0.6576	0.7689	0.6797
7	0.7186	0.6662	0.7084
8	0.7384	0.7506	0.7532
9	0.775	0.6751	0.8187
10	0.7683	0.7239	0.768

Table 3 Levenberg-Marquardt backpropagation model selection for second case (MSE)

No. of nodes	No. of delay		
	2	3	4
2	0.0055	0.014	0.0133
3	0.0075	0.0236	0.0065
4	0.0089	0.0108	0.0074
5	0.0189	0.0051	0.0073
6	0.0144	0.0071	0.0136
7	0.6596	0.7923	0.0192
8	0.1429	**0.0041**	0.0204
9	0.017	0.0144	0.0156
10	0.0124	0.0101	0.0253

Table 4 Levenberg-Marquardt backpropagation model selection for second case (*R*)

No. of nodes	No. of delay		
	2	3	4
2	0.8269	0.6404	0.6882
3	0.8003	0.6136	0.786
4	0.716	0.73	0.8661
5	0.7017	0.8339	0.7268
6	0.491	0.8277	0.5985
7	0.7817	0.78025	0.6215
8	0.5573	**0.8451**	0.622
9	0.597	0.6466	0.7237
10	0.6427	0.7394	0.6331

Table 5 Levenberg-Marquardt backpropagation model selection for third case (MSE)

No. of nodes	No. of delay		
	2	3	4
2	0.0117	0.0074	0.0121
3	0.9863	0.0021	0.0101
4	0.0063	0.0077	**0.002**
5	0.9092	0.0062	0.0197
6	0.0067	0.0183	0.0103
7	0.019	0.0133	0.0101
8	0.0129	0.0077	0.0113
9	0.0125	0.0062	0.0135
10	0.0184	0.0058	0.0218

Table 6 Levenberg-Marquardt backpropagation model selection for third case (*R*)

No. of nodes	No. of delay		
	2	3	4
2	0.7489	0.7989	0.7048
3	0.7042	0.7226	0.6843
4	0.7912	0.7989	**0.8302**
5	0.7033	0.6993	0.5561
6	0.8	0.6217	0.7185
7	0.643	0.6912	0.859
8	0.5721	0.7396	0.58
9	0.63	0.7889	0.666
10	0.5742	0.7971	0.6566

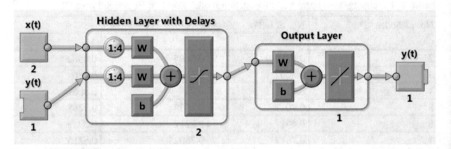

Fig. 3 Neural network of third case

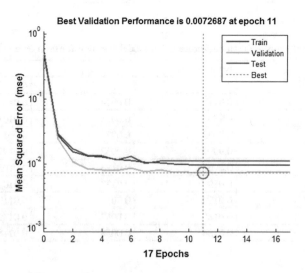

Fig. 4 Performance plot

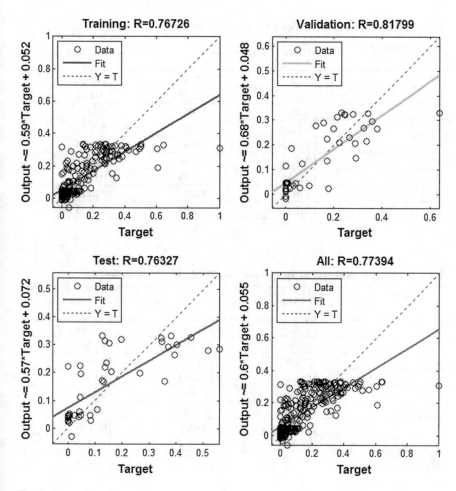

Fig. 5 Regression plots

The learning rate 0.005 is used in Bayesian regulation backpropagation algorithm to train the data. The experiment on the model was repeated by using the number of nodes from 2 to 10 and number of tapped delays from 2 to 4. The MSE value was observed for all networks of all three cases.

From Tables 7, 8, 9, 10, 11 and 12, it is clear that the lowest MSE value is 0.0032 with five nodes in hidden layer and four tapped delay lines (first case), 0.0079 with two nodes in hidden layer and two tapped delay lines (second case) and 0.0061 with four nodes in hidden layer and four tapped delay lines (third case). Lowest MSE is supported by high value of R value 0.8277, 0.7438 and 0.8563, respectively.

Table 7 Bayesian regulation backpropagation model selection for first case (MSE)

No. of nodes	No. of delay		
	2	3	4
2	0.0077	0.0173	0.0094
3	0.0093	0.0129	0.0089
4	0.0109	0.005	0.0156
5	0.0109	0.0108	**0.0032**
6	0.0125	0.0068	0.0054
7	0.0118	0.0049	0.007
8	0.0138	0.0118	0.018
9	0.0128	0.006	0.0115
10	0.0121	0.0152	0.0113

Table 8 Bayesian regulation backpropagation model selection for first case (R)

No. of nodes	No. of delay		
	2	3	4
2	0.7621	0.671	0.7512
3	0.7367	0.6186	0.8308
4	0.6852	0.8259	0.7812
5	0.7244	0.6712	**0.8471**
6	0.774	0.803	0.8314
7	0.6744	0.8536	0.8351
8	0.648	0.6524	0.7076
9	0.655	0.8034	0.7225
10	0.7931	0.7701	0.7733

Table 9 Bayesian regulation backpropagation model selection for second case (MSE)

No. of nodes	No. of delay		
	2	3	4
2	0.011	**0.0079**	0.0142
3	0.0242	0.0193	0.0201
4	0.0186	0.013	0.0199
5	0.0175	0.0088	0.0286
6	0.0114	0.0235	0.0191
7	0.022	0.0194	0.0158
8	0.0227	0.0123	0.0125
9	0.0108	0.0107	0.0095
10	0.0125	0.0097	0.0156

Table 10 Bayesian regulation backpropagation model selection for second case (R)

No. of nodes	No. of delay		
	2	3	4
2	0.735	**0.7438**	0.641
3	0.681	0.7561	0.6546
4	0.6156	0.6995	0.6056
5	0.6731	0.7517	0.6696
6	0.6572	0.5715	0.7276
7	0.7866	0.576	0.682
8	0.6372	0.6091	0.6522
9	0.7502	0.7076	0.698
10	0.6849	0.6766	0.7931

Table 11 Bayesian regulation backpropagation model selection for third case (MSE)

No. of nodes	No. of delay		
	2	3	4
2	0.0108	0.0086	0.0116
3	0.0078	0.0087	0.0082
4	0.0169	0.0079	**0.0061**
5	0.0104	0.0092	0.0165
6	0.0116	0.0071	0.0176
7	0.0141	0.0077	0.0073
8	0.0343	0.0117	0.0083
9	0.01	0.0088	0.0102
10	0.0084	0.0087	0.0094

Table 12 Bayesian regulation backpropagation model selection for third case (R)

No. of nodes	No. of delay		
	2	3	4
2	0.791	0.7441	0.6591
3	0.7844	0.7663	0.7401
4	0.6764	0.8239	**0.8563**
5	0.71	0.8347	0.6343
6	0.7264	0.777	0.7349
7	0.6489	0.7776	0.8311
8	0.4908	0.7335	0.768
9	0.7577	0.7786	0.7731
10	0.7594	0.7788	0.7101

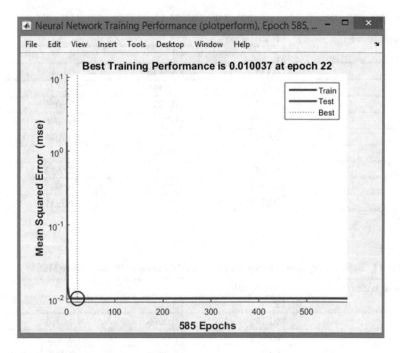

Fig. 6 Performance plot

MSE graph of neural network model produced during training by Bayesian regulation backpropagation training algorithm is shown in Fig. 6. The training steps ended at the end of seventeenth epochs as the validation errors increased. At the end of neural network model training, MSE (mean square error) is obtained as 0.0072687 as shown in figure above, whereas 'R' values are above 0.76. This implies that the output produced by the network is closer to the target and that the model is suitable (Fig. 7).

5 Conclusion

To predict the rainfall (in mm) in Burdwan, the best model was selected from both the algorithms using MSE of all cases as performance indicator.

	First case		Second case		Third case	
	LM	BP	LM	BP	LM	BP
MSE	0.006	**0.0032**	**0.0041**	0.0079	**0.002**	0.0061
R	0.8277	0.8471	0.8451	0.7438	0.8302	0.8563

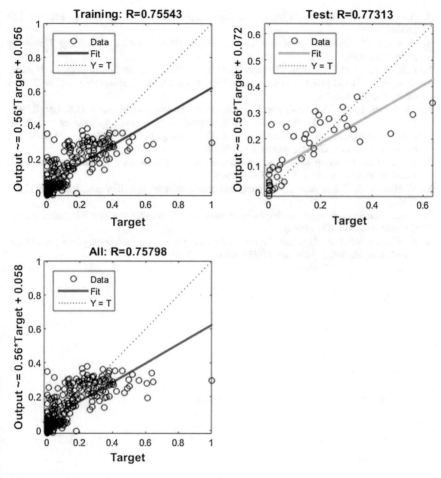

Fig. 7 Regression plots

The lower value of MSE in first case is given by Bayesian regulation back-propagation algorithm and in second and third cases is given by Levenberg-Marquardt backpropagation algorithm.

References

1. S. Abudu, J.P. King, A.S. Bawazir, Forecasting monthly streamflow of spring-summer runoff season in Rio Grande headwaters basin using stochastic hybrid modeling approach. J. Hydrol. Eng. **16**(4), 384–390 (2011)
2. S. Chen, S.A. Billings, P.M. Grant, Non-linear system identification using neural networks. Int. J. Contr. **51**(6), 1191–1214 (1990)

3. A.J. Jones, New tools in non-linear modeling and prediction. Comput. Manag. Sci. 109–149 (2004)
4. H. Kantz, T. Schreiber, *Nonlinear Time Series Analysis*, 2nd edn. (Cambridge University Press, Cambridge, 2006)
5. R.K. Pachauri et al., Climate change 2014: synthesis report, in *Contribution of Working Groups I, II and III to the Fifth Assessment Report of the Intergovernmental Panel on Climate Change (IPCC)* (2014)
6. J.R. Porter, L. Xie, A.J. Challinor, K. Cochrane, M. Howden, M.M. Iqbal, D.B. Lobell, M.I. Travasso, Chapter 7: Food security and food production systems. Climate change 2014: impacts, adaptation and vulnerability, in *Working Group II Contribution to the IPCC 5th Assessment Report, Geneva, Switzerland* (2014)
7. T. Lin et al., Learning long-term dependencies in NARX recurrent neural networks. IEEE Trans. Neural Netw. 7(6), 1329–1338 (1996)
8. Y. Hao, B.M. Wilamowski, Levenberg–marquardt training. Ind. Electron. Handb. 5(12), 1 (2011)
9. D.J.C. MacKay, A practical Bayesian framework for backpropagation networks. Neural Comput. 4.3, 448–472 (1992)
10. A. Khamis, S.N.S.B. Abdullah, Forecasting wheat price using backpropagation and NARX neural network. Int. J. Eng. Sci. (IJES) 3, 19–26 (2014)

Computational Intelligence
via Applications

Parth Mahajan and Kushagra Agrawal

Abstract The constant mobile revolution and the increasing urge for smartphone usage are really hiking up the market for mobile applications. The growth in the customer demand for mobile applications is rapidly increasing. According to the research report, there are more than 102,062 million apps downloaded globally presently. It is expected rise to 268,692 million by 2017. As the mobile applications are getting highly adaptive, the best way to make the artificial or computational reach the people is through mobile applications. This can be easily achieved as we can keep a larger amount of knowledge base on the servers, and it can be further accessed by the applications via Internet. In this paper present a simple and a open source approach for this by making an application possessing artificial intelligence, which is globally being used as "G-Connect".

Keywords Ubiquitous access · Computational intelligence · IDE (Integrated Development Environment) · Knowledge base · Mobile and web applications

1 Introduction

The connotation of the word computational intelligence can be depicted as the ability of any software, machine, program or application to learn specific tasks and actions from the data/knowledge base provided or from the experimental observations. It is a set of nature-inspired computational methods which very closely replicates the humans' way of thinking. Thus, these revolutionary methodologies help placating real-life problematic situations and also amortising extensive human effort [1, 2].

P. Mahajan (✉) · K. Agrawal
GD Goenka University, Sohna Road, Gurgaon 122103, Haryana, India
e-mail: parthmahajan1310@gmail.com

K. Agrawal
e-mail: kushagra.agarwal1@gdgoenka.ac.in

© Springer Nature Singapore Pte Ltd. 2018
B. Panda et al. (eds.), *Innovations in Computational Intelligence*, Studies in
Computational Intelligence 713, https://doi.org/10.1007/978-981-10-4555-4_20

This paper first provides information about the type of applications that can be constructed using various technologies. Further are followed by some examples of existing popular applications possessing artificial intelligence. It will also explain how simple the artificial intelligence can be implemented within the applications by a simple experiment. From this paper, we will also get a hunch of what all have been already implemented and can be possibly be achieved from AI if it uniformly blends up with mobile application.

Present paper presents a portable, free of cost and a ubiquitous native application. Present article focuses on how artificial intelligence can be spread among the maximum of population by creating right type of applications. Keeping in mind the target customers and the specific use cases to be provided.

1.1   Which Type of Application to Go for: Native, Hybrid or Web App

Before we dive into the development of the application, we must be clear which type of application will be the most suited one in currently prevailing scenario. Which one is the most appropriate can be decided on the basis of our use case and the factors like your budget, timescale and priority [3, 4].

Native Apps: The native apps are typically the type of application we mostly use. They are available on the official app store of the company. They are platform dependent that means they are designed and coded separately for each platform, e.g. for android, the apps are developed in android studio or eclipse using java. Similarly for iOS, the apps are developed using objective C or swift. Each platform provides developers with its own kind of standardised Software Development Kit (SDK). They have the fastest, most reliable and most responsive user interface.

Examples: Angry Birds, Shazam [5, 6].

Web Apps: We all must have seen the "mobile version" of a website, those are nothing but the web applications intimating that website. The web app like other websites loads into the mobile browser like Chrome or Safari. They are ideal to the native app in terms of design, look and feel. The advantage here is we do not have to download the app on to our device and avail space of it. However, its development is facile and nimble, but the developers cannot access the standardised SDKs in this case. Also they have limited access to hardware and provide less features than the other 2 [5].

Hybrid Apps: The hybrid apps lie between native and web apps. They are faster and easier to build than native apps and use cross-platform web technologies to build such as HTML, CSS and java script. They are easier to maintain as they are cross-platform. Also they can interact with the hardware and can provide the similar kind of functionalists as that of native apps through different plug-ins provided. But it takes substantial work to run them appropriately and seamlessly on different platforms. The performance is very close to native apps but still lags a bit in terms of response time over the Internet. It is also easier for the developers to replicate any newly introduced feature over all the other platforms.

Examples: Facebook, LinkedIn, Twitter and Amazon app store [7].

1.2 Existing Applications Possessing AI

We are provided with a bunch of applications which have artificial intelligence. A few of them I have mentioned below.

Siri: It is a virtual assistant which have a voice-controlled natural language interface. It also helps iOS users to perform their daily tasks like playing some music, opening gallery, texting or calling someone and many more.

AI Webcam: It is an application which surveillance and home security using artificial intelligence. It just requires a simple web camera, rest is managed by the app itself. The application can overwhelm the human efforts as it can work as a hidden camera possessing features such as instant notification, motion detection, hiding ability security and much more [7, 8].

Cortana: It is an application provided my Microsoft. It is integrated in Windows 10 and is provided cross-platform as well. It has the ability to text people, make a call, tell jokes, open calendar, write notes and many more.

There are many such applications which offer similar kind of features like Google now, Assistant, AIVC, Jarvis and Hound.

All these advanced applications mentioned above are frequently used by the application users and that can be easily noticed by having a look at the no. of downloads of each of them on their respected app stores. All of them possess some of the AI features mentioned below:

1. Optical character recognition
2. Handwriting recognition
3. Speech recognition
4. Face recognition
5. Artificial creativity
6. Pattern recognition

1.3 About My Application "G-Connect"

This android application is one-stop interface providing all the courses offered and their related contents, keeping students updated about the latest events and announcements. This application also provides complete information about GD Goenka University and its admission details for aspiring students.

Fig. 1 Admission module of
the application "G-Connect"

An android application for GD Goenka University to bring contents of varied courses, admission-related details and latest events and announcements on one single platform. It consists of following modules:

1. About GDGU: Detailed information about GD Goenka University
2. Admission: The interested students can fill the enquiry form and submit it.
3. Study Material: The existing students can login and get access to features like as follows:
   (a) Lecture notes (ppts and pdfs)
   (b) Syllabus
   (c) Assignments
   (d) Timetable
4. News and Events: News about the latest events held in the university can be viewed and accessed.
5. Library: Availability of books in the library can be viewed.
6. Notice board/Departmental Updates: All the updates and announcements for the students.
7. Academic calendar for the current semester.

Fig. 2 Home page with
variety of options to select
from

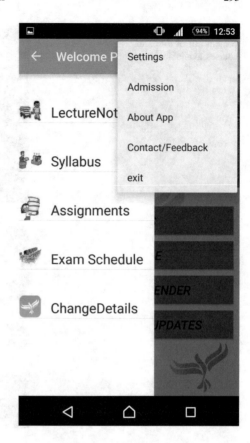

8. Location: Location on the map where the university is situated.
9. Contact us: E-mails and contact numbers to contact university for any query
   (Figs. 1 and 2)

2 Experiment

In this paper, I have presented a native type of android application. The application
stores all types of information in a SQLite database and uses that as its knowledge
base. Then, on the basis of logical programming, it produces artificially intelligent
results by using this knowledge base (Figs. 3 and 4).

2.1 Open Source and Cost Free IDE

Android Studio: This IDE (integrated development environment) as it the official
IDE provided by the Google. It provides the most appropriate and the fastest tools

Fig. 3 Selecting required content out of different courses available

to build applications for every type of android devices. This IDE is freely available. It is available most of the operating systems including Mac OS X, Windows and Linux. It has also replaced Eclipse ADT (android development tools) and has become the primary IDE for developing native apps for android.

Bluestacks: It is an emulator which gives us an option for putting android on our PC. With the help of Bluestacks, we can install our application and also test our application on the PC. As it is not possible to have android phones of different screen sizes and different versions of the operating system, we can be sure that the application does not harm our running operating system on the mobile phone. It is a one-stop solution for all these problems (Fig. 5).

2.2 Working

The present application is based on intelligently retrieving the suitable and correct information from the given knowledge base through a simple, attractive and easy to use graphical user interface.

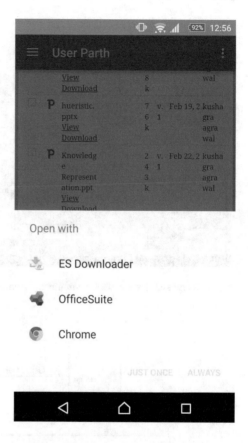

Fig. 4 Different options for the content to be viewed or downloaded

When the application starts, a welcome page is shown for 1.5 s. Meanwhile it checks whether the user has already logged in to the application or not. If yes, then it is directed to the home page of the application with his username on the top, otherwise he is taken to a page from where he can either jump to the sign up page or to admission to admissions page. On the university page, he will find each and every detail about the university in detail such as different courses and schools, news and events, media, feedback form and much more.

On the other hand, if he switches to the sign up page, then he will come across a form where he has to provide his basic details like course, semester and username. This information is stored in a SQLite database. So, on the reopening of the application, we will be taken to the home page directly with only the content which is relevant for us. This database acts as a knowledge base here from which the information is retrieved using computational intelligence.

Now as we proceed and select different options for accessing different contents, we are taken to new activities that have suitable content and image views. Further, we can either download the study material or view it within the application. Timetable like these are provided offline within to save the data usage (Fig. 6).

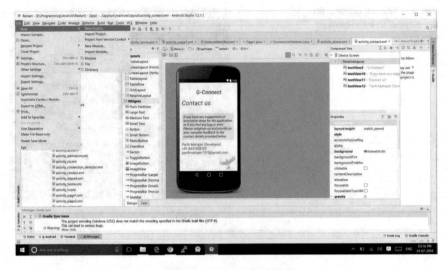

Fig. 5 Integrated development environment "Android Studio"

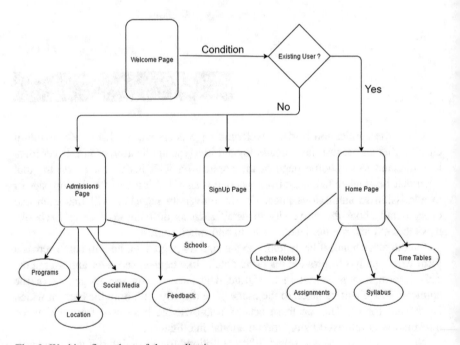

Fig. 6 Working flow chart of the application

3 Conclusion

Apps play a vital role in day-to-day life and for users it acts as a tool box, which leads users towards the huge and emerging world of technology. So we can conclude these applications are highly engaged in people's life, and also it is the best way out for taking the artificial intelligence to the majority of people. These apps possessing AI have a strong potential to handle upcoming needs of people worldwide and can serve as a complete personal assistant and can lower down tremendous amount of human effort, making world a better place.

4 Future Direction

Our future and ultimate objective is the ability to create applications which can learn on their own by observing the surroundings and learning from past experience. They must possess the ability to take right decisions on their own, and they also must have the common sense to manipulate their decisions according to constantly changing situations and automatically deducing itself for immediate consequences of anything it is being told on the basis of what is already known.

References

1. S. Palaniappan, N. Hariharan, N.T. Kesh, S. Vidhyalakshimi, A. Deborah, Home automation systems—a study. Int. J. Comput. Appl. **116**(11) (2015) (0975 8887)
2. G. Jindal, M. Anand, Android applications with artificial intelligence in mobile phones. Res. Expo. Int. Multidisc. Res. J. **II**(IV). Available online at www.researchjournals.in (2012). ISSN: 2250-163
3. Accenture Digital, *Growing the Digital Business: Spotlight on Mobile Apps* (Accenture Mobility Research, 2015)
4. M. Shiraz, A. Gani, R. Hafeez Khokhar, Member IEEE, R. Buyya, A review on distributed application processing frameworks in smart mobile devices for mobile cloud computing, IEEE Commun. Surv. Tutorials, **15**(3) (THIRD QUARTER 2013)
5. W. Jobe, *Native Apps vs. Mobile Web Apps* (Stockholm University, Stockholm, Sweden)
6. D. Chaffey, *Mobile Marketing Statistics Compilation* (Mobile Marketing Analytics, April 27, 2016)
7. L. Ma, L. Gu, J. Wang, Research and development of mobile application for android platform. Int. J. Multimedia Ubiquit. Eng. **9**(4) (2014)
8. C. Kaul, S. Verma, A review paper on cross platform mobile application development IDE. IOSR J. Comput. Eng. (IOSR-JCE) **17**(1), 30–33 Ver. VI (2015)

Web Page Representation Using Backtracking with Multidimensional Database for Small Screen Terminals

Shefali Singhal and Neha Garg

Abstract Nowadays Web plays a vital role in everyone's life. Whether one has to search something, perform any online task or anything on Internet, one has to come across Web pages throughout. For small screen terminals such as small mobile phone screen, palmtop, pagers, tablet computers, and many more, reading a Web page is time-consuming task because there are many unwanted content parts which are to be scrolled or there are advertisements, etc. To overcome such issue, the source code of a Web page is organized in tree format. In this paper, backtracking is applied on the tree as per some mentioned rules (I, II, III) to filter out the Web page headings. These rules are applied until none of the nodes is left unvisited. The filtered data is then mapped to a multidimensional (MDB) data cube. Again MDB is used by online analytical processing (OLAP) which in turn is one of the greatest tools for analysts in today's entrepreneur's scenario. This strategy makes Web page access a very simple and interesting task.

Keywords DOM · Block extraction · Backtracking · Data cube

1 Introduction

In this era, a Web page is a very useful entity for both the end-user and the developer in different fruitful ways. But as a developer, it has been always promoted to make the things simpler and easier for end-user. To accomplish this goal, one can use various enhancement techniques, algorithms, designing of interface, and many more things.

S. Singhal (✉) · N. Garg
CSE Department, Manav Rachna International University, Faridabad Sec-43,
Aravalli Hills, Faridabad 121006, India
e-mail: shefali.fet@mriu.edu.in

N. Garg
e-mail: nehagarg.fet@mriu.edu.in

© Springer Nature Singapore Pte Ltd. 2018
B. Panda et al. (eds.), *Innovations in Computational Intelligence*, Studies in
Computational Intelligence 713, https://doi.org/10.1007/978-981-10-4555-4_21

In previous paper, from a Web page independent blocks were created using hybrid of DOM (document object model) and VIPS (vision-based page segmentation) techniques. These blocks contain the textual part of a heading in a Web page. For an end-user, heading text is the most useful entity of a Web page comparatively.

In this paper, the above-discussed blocks, along with various other information under a heading of a Web page, are separately stored using cube data structures (OLAP model of data mining) and presented on user interface accordingly.

This paper is organized as follows: The related work is explained in the section Literature Review. Then, we have discussed the technique of *Block Extraction*, which we have presented in our previous paper [1]. In next section, *Backtracking Technique to Find Headings* with the help rules and example is explained. Further, the method to represent data in *Multidimensional Cube* is explained. Finally, we draw conclusion and present our future work in *Conclusion and Future Work* section.

2   Literature Review

DOM and block extraction [1] are techniques which provide hierarchical structure of the source code, i.e., HTML code of a Web page. Then, there is concept of backtracking for this tree. Then, a data cube is created.

One more concept has been proposed [5], where a vision-based page segmentation algorithm is used to partition a Web page into semantic blocks with a hierarchical structure. Then, spatial features (such as position and size) and content features (such as the number of images and links) are extracted to construct a feature vector for each block. Based on these features, learning algorithms are used to train a model to assign importance to different segments in the Web page.

Another concept is [8] that ECON finds a snippet-node by which a part of the content of news is wrapped firstly, and then backtracks from the snippet-node until a summary-node is found, and the entire content of news is wrapped by the summary-node. During the process of backtracking, ECON removes noise. It was also very effective related work.

For knowledge of OLAP and data cube [13], the basic logical model for OLAP applications has been proposed. Further created two subcategories: the relational model extensions and the cube-oriented approaches, which are really impactful.

3   Block Extraction

Block extraction algorithm has been proposed earlier [1]. This algorithm is applied on a DOM tree (which is created from source code of a Web page). Each node of DOM tree is visited sequentially. Few conditions are mentioned to be fulfilled by a node:

(1) Divide a node if the DOM node has only one valid child and the child is not a text node.
(2) Divide a DOM node if one of the child nodes of the DOM node is line-break node.
(3) Divide a DOM node if one of the child nodes of the DOM node has HTML tag <HR>.
(4) Divide a DOM node if the background color of this node is different from one of its children, but the child node with different background color will not be divided in this round [1].

After applying these conditions, nodes can be categorized as blocks. A block is a leaf of resulting tree. The content of a block is core text content under a heading of a Web page. These blocks are independent. These blocks are further arranged into a queue which is pool of blocks. Thus, Web page segmentation concept provides a new mode for data arrangement and also parsed data content.

Now, one can use these blocks for any further research purpose.

4   Heading Search Through Backtracking

Till now, we are having DOM tree, block containing tree, and independent blocks. Backtracking is to be applied on DOM tree.

What is backtracking?

Backtracking is simply a tree traversal algorithm in which a particular entity is searched on the basis of some given condition. With respect to tree traversal, backtracking is simply visiting a tree starting from its bottom and moving toward top of the tree. For example, starting from a leaf and then jumping to previous internal nodes on same path.

The DOM tree contains various HTML tags in its nodes. Here, our interest is in finding main headings of the Web page.

Rule I
How to find main headings?

1. Visit a leaf node in tree.
2. Start backtracking and traverse the tree from bottom to top.
3. If an internal node contains heading tag (<H1>), then again backtrack once for node just above heading node.
4. If the node found in step 3 is division tag (<div>), then it means the encountered heading node is actually a subheading, so continue backtracking till a node with body tag (<body>) is encountered.
5. If body tag is found, move down once and reach heading tag which is now justified as one of the main headings of the Web page.

Rule II

Following are few rules to be applied when a main heading node is found in a tree:

1. Stop backtracking.
2. Place heading into an array which is to be used for heading collection.
3. Start visiting sibling nodes.
4. Declare a queue for this heading.
5. Place sibling into queue if it is image, hyperlink, or text.
6. If the sibling node contains "div," then make it a reference node and follow Rule III.
7. Go to another leaf node for finding next main heading in Web page.

Rule III

Following are few rules to be applied when a subheading node is found in a tree:

1. Continuing from step 6 of Rule II, a sibling node containing division tag is referenced.
2. Declare another array for subheadings of the particular main heading found previously.
3. Connect the above array with division node by referencing.
4. Proceed as per step 4 to step 7 of Rule II.

For multiple levels of subheadings, one can apply same methodology again level-wise.

Flowchart for Rules I, II, III (Fig. 1)

Example of backtracking as per Rules.
In the DOM tree (Fig. 2), leaves are visited sequentially using backtracking technique. One has to visit a path from bottom to top and filter the required data. For example, suppose leaf node "H1" is visited. On applying Rule I, next internal node is "body." It means "H1" is one of the main headings of the Web page. Now, once it is decided that current node is a main heading node, one must traverse the siblings of current node as per Rule II. All the siblings are added into a queue except div node which is further traversed as per Rule III. Thus, one can conclude how main headings are listed and ready for small screen terminals.

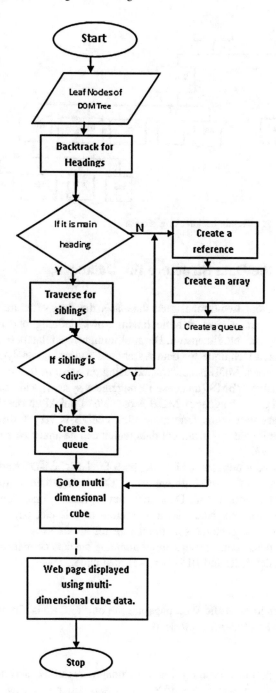

Fig. 1 Flowchart of the complete process

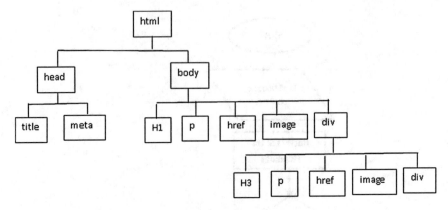

Fig. 2 DOM tree to represent backtracking process

5   Cube as the Data Structure for Database

When we talk about two-dimensional database, the data structure is a relational database (RDB), i.e., simply a table having rows defining one dimension and columns defining second dimension. But multidimensional database (MDB) proves to be an optimized technique for data warehouse and online analytical processing (OLAP) applications. MDB is constructed using an existing RDB; however, RDB is constructed using DBMS (database management system) and accessed through SQL (structured query language). MDB is capable of quickly processing any query and replying very efficiently. One more thing, MDB is very fruitful for analysis purpose because it contains historical data which can be analyzed and some results can be jotted down.

Data cube is very popular, which can be 3-D, 4-D,…64-D,..And so on. It represents different entities on different axes. One can utilize it for many more dimensions as per requirement. Data cube can represent dimensions of data contained with the user. For example, a data cube of 6-D can have sales, product, region, quantity, cost, customers as the names for all six axes.

Here, the proposed concept says about mapping the data contained in queues and arrays as per Rules I, II, and III into a 3-D data cube.

X-AXIS

All the main headings of the Web page are the unit for x-axis. So, it will show all the array elements of step 2 of Rule II.

Y-AXIS

All the content part of a heading, i.e., text, image, hyperlink, and division will be displayed as unit on y-axis. So, if x- and y-axis are mapped at some point, a combination of main heading and its content will be resulting.

Fig. 3 Multidimensional database representation for a Web page

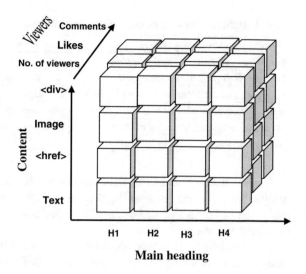

Main heading

Z-AXIS

All the responses of viewers of Web page, i.e., comments, number of visitors, and likes will be displayed as unit on z-axis (Fig. 3).

Now, if a point is plotted on graph where all the three axes are mapped as per above description, then one can find a heading followed by its content including all subheadings and viewers which is actually a complete heading in itself. The same is resulting if above figure is analyzed.

6 Heading Extraction for Small Screen Terminals

After block extraction and data cube creation, our main aim is to organize these elements in such a manner so that any GUI on small screen terminal (SST) can be much simpler and easier to use/access. So, if a normal Web page is accessed on SST, then there is a problem of scrolling the page either height-wise or width-wise again and again for every heading. Also, there are various advertisements and pop-ups which are really interrupting for a user because of which accessing actual information is a time-consuming and tough task.

To overcome all these problems, a concept is proposed in which we will be using all the aspects collected till now. For SST, a Web page will be showing only list of main headings it contains. The user will go through those headings and select a particular heading in which he is interested. Once a heading is selected, the user will get consecutively all of its contents and subheadings in which he is interested. This will lead to a proper view of all the Web page content without navigation problem in SST.

Table 1 Difference between existing and proposed technique

Ongoing common access method	Proposed method
Cannot read all headings in single view	Read the heading list completely
Navigate Web page as per screen size	Not much navigation due to listing
Bounded to search more for relevant heading	Go only for relevant heading
Read all content in relevant heading	Skip uninterested content within a heading
Inefficient and time-consuming method	Time saver and quick responding method
No such database supporting analysis	Data cube is meant for analytical processing

One more effective use of data cube is for the businessmen or higher level analyst who can get support in decision making which may further lead to expansion or contraction of business level. So, this strategy is a boon for both the end-users and Web site owner.

If comparison is made between current concept and what actually exist, there are many aspects in which the proposed concept will sustain as the optimized one (Table 1).

7 Conclusion and Future Work

This paper is in continuation with [1]. So, starting from the extracted blocks one more concept is proposed which talks about application of backtracking on DOM tree and proceeding further by following Rules I, II, and III in this paper. Once completed with tree traversal, data cube comes in picture in which the extracted content trough backtracking is organized which is in the form of three-dimensional database. The axis with heading as unit is displayed on user interface which proves to be enhanced Web page navigation for SST.

There is a great scope as per future perspective for the above-mentioned entities. The data cube constructed in this paper is independent in itself which can be used further in cloud computing, Web semantics, advanced data mining, knowledge discovery, information retrieval, and many more computer research fields. This technique for the frontend of Web page, i.e., user interface can be applied by various Web page developers for SST. So, this work can be extended in multiple directions for further research.

References

1. S. Shefali, N. Garg, Hybrid web-page segmentation and block extraction for small screen terminals, in *IJCA Proceedings on 4th International IT Summit Confluence 2013—The Next Generation Information Technology Summit Confluence* (2013) (2), 12–15 Jan 2014

2. M. Álvarez, A. Pan, J. Raposo, F. Bellas, F. Cacheda, Extracting lists of data records from semi-structured web pages, Department of Information and Communication Technologies, University of A Coruña, Campus de Elviña s/n, 15071 A Coruña, Spain, 11 Oct 2007
3. D. Cai, S. Yu, J.-R. Wen, W.-Y. Ma, VIPS: a vision-based page segmentation algorithm, in *Microsoft Research Microsoft Corporation One Microsoft Way Redmond*, WA 98052, Nov 1 2003
4. S. Baluja, Browsing on small screens: recasting web-page segmentation into an efficient machine learning framework, in *Proceeding of the 15th International Conference on World Wide Web*, 2006-05-23
5. R. Song, H. Liu, J.-R. Wen, W.-Y. Ma, Learning block importance models for web pages, in *Proceedings of the 13th International Conference on World Wide Web*, 2004-05-17. ISBN:1-58113-844-X
6. D. Cai, S. Yu, J.-R. Wen, W.-Y. Ma, Block-based web search, in *Proceedings of the 27th Annual International ACM SIGIR Conference on Research and development in Information Retrieval*, 2004-07-25
7. K. Vieira, S. Altigran da Silva, N. Pinto, S. Edleno de Moura, J.M.B. Cavalcanti, J. Freire, A fast and robust method for web page template detection and removal, in *Proceedings of the 15th ACM International Conference on Information and Knowledge Management*, 2006-11-06
8. Y. Guo, H. Tang; L. Song, Y. Wang, G. Ding, ECON: an approach to extract content from web news page, in *2010 12th International Asia-Pacific Web Conference (APWEB)*, 6–8 Apr 2010
9. F. Zhao, The algorithm analyses and design about the subjective test online basing on the DOM tree, in *2008 International Conference on Computer Science and Software Engineering*, 12–14 Dec 2008
10. G. Colliat, OLAP, relational, and multidimensional database systems. ACM SIGMOD Record **25**(3) (1996)
11. P. Vassiliadis, Modeling multidimensional databases, cubes and cube operations, in *Proceedings of Tenth International Conference on Scientific and Statistical Database Management*, 3 July 1998
12. C. Stolte, D. Tang, P. Hanrahan, Polaris: a system for query, analysis, and visualization of multidimensional relational databases. IEEE Trans. Vis. Comput. Graph. (Jan/Mar 2002) **8**(1) (2002)
13. P. Vassiliadis, T. Sellis, A survey of logical models for OLAP databases. ACM SIGMOD Record **28**(4) (1999)

Printed in the United States
By Bookmasters